Deliberate Coaching

Optimizing Teaching and Learning Through Behavior Science

EDUCATION EDITION

Deliberate Coaching

Optimizing Teaching and Learning Through Behavior Science

EDUCATION EDITION

Paul Gavoni and Nicholas L. Weatherly

KeyPress
Publishing

KeyPress Publishing
www.keypresspublishing.com

 KeyPress
Publishing

Authors: Paul Gavoni and Nicholas L. Weatherly

Deliberate Coaching: Optimizing Teaching and Learning Through Behavior Science

Published by: KeyPress Publishing
Publisher: Alice Darnell Lattal
Brand Integrity and Design: Jana Burtner
Production Manager: Adele Hall
Editors: Mary Sproles Martin, Ashley Johnson, and Stefanie Carr

ISBN 979-8-9886548-6-5

Library of Congress Control Number: 2024943033

Published in Melbourne, Florida

Distributed by:
ABA Technologies, Inc.
930 South Harbor City Blvd, Suite 402
Melbourne, FL 32901
www.abatechnologies.com

KeyPress Publishing books are available at a special discount for bulk purchases by corporations, institutions, and other organizations. For more information, please email keypress@abatechnologies.com.

Table of Contents

Foreword

I've been at the helm of both center-based and school-based special education for over a decade now, and let me tell you, nothing beats seeing the impact of solid training and feedback on our educators firsthand. Leading the charge in an alternative school throws unique curveballs your way, requiring a keen grasp of coaching methods that truly inspire growth and innovation. As a principal, I've steered a wildly diverse team of educators and students, always on the lookout for new ways to unlock potential and drive real change. The ideas in *Deliberate Coaching* have been a game changer for my leadership style and have seriously revived our school community.

Dr. Paul "Paulie" Gavoni, or Paulie as I know him, has been a powerhouse in championing a values-driven approach in education, all woven together with Applied Behavior Analysis (ABA). His magic lies in sparking performance upgrades wherever he goes by setting up killer systems, rolling with servant leadership, and nailing it with his coaching style based on the good stuff—positive reinforcement.

Deliberate Coaching, penned by Dr. Gavoni and another powerhouse in Organizational Behavior Management (OBM), Dr. Nic Weatherly, dishes out a systematic, science-backed road map that empowers school leaders to reach their goals. With a focus on intentional coaching rooted in behavioral science, this book is a treasure trove for any school leader aiming to elevate their school's vibe, beef up teaching quality, and boost student success.

In the world of OBM, *Deliberate Coaching*, in my opinion, is breaking new ground. The authors dive into the gnarly issues in education without flinching, armed with a sharp, solution-focused mindset that's rooted deep in behavioral science. Their influence isn't just local; it stretches across the globe through podcasts and top-selling books. Their philosophy is all about pushing for constant growth and nailing practical effectiveness, setting *Deliberate Coaching* up as a true model of integrity and excellence in education.

Diving into the principles of human behavior science and coaching strategies detailed in *Deliberate Coaching* is critical, especially for those of us in the trenches of alternative schools dealing with students facing tough challenges. These schools are no cakewalk, often filled with students who've had a rough ride in typical educational settings due to behavioral issues, academic bumps, or other hurdles.

Deliberate Coaching lays out a solid blueprint for leaders navigating these complex school scenarios. By spotlighting targeted interventions, personalized support, and a structured coaching approach, the book arms leaders with actionable strategies to guide educators in managing tough behaviors effectively. These methods not only boost teacher confidence but also ensure they're equipped to meet the varied needs of our students.

Moreover, the book champions a culture of collaboration, trust, and ongoing improvement within school communities. Creating a supportive atmosphere is pivotal to success in schools, especially alternative and Title I schools, where challenges can be steep and resources slim. Through the coaching practices outlined, school leaders and those supporting school improvement efforts can foster a team spirit and shared responsibility among educators, boosting their ability to handle even the most demanding behaviors.

Plus, *Deliberate Coaching* highlights the vital role of continuous professional development and feedback in promoting teacher growth. In the dynamic and often intense world of education, it's essential for educators to keep learning. This book offers insights into providing constructive feedback, setting clear goals, and implementing targeted interventions to positively shape educators' performance.

In essence, the principles of *Deliberate Coaching* are key to creating safer and more effective learning environments for students in schools. By equipping educational leaders with the right tools and strategies, this book helps schools cater to the diverse needs of their students and initiate meaningful change. In the often-challenging landscape of education, *Deliberate Coaching* is an invaluable asset for any leader looking to foster growth, innovation, and positive outcomes for students.

A major takeaway from this book is the importance of unwavering support right from the start. As a school leader, your role in backing your educators from day one, from the initial job interview to ongoing classroom evaluations, is critical in fostering a culture of support and coaching that's key to nurturing teacher development and, in turn, boosting student outcomes.

But remember, guiding teachers to success isn't just a one-off thing—it's about being all in, all the time. By focusing on boosting teacher effectiveness through regular and Deliberate Coaching, school leaders can accelerate major improvements in student achievement. This requires fostering a campus-wide culture marked by collaboration, trust, and continuous learning.

Deliberate Coaching at the individual level goes beyond usual tactics by pinpointing specific behaviors and skills of educators within our strategic communication and processes. Through personalized Deliberate Coaching at my school, we've helped teachers tackle hurdles and harness their strengths, significantly contributing to an atmosphere of excellence that permeates every aspect of our school community. Even the union is happy with our approaches!

Effective coaching transcends individual performance. It involves developing and implementing systems that support effective teaching and behavior management across the school. In our alternative school setting, our Deliberate Coaching approaches drive sustainable change. By mastering the art and science of focusing on individual actions while also considering the bigger organizational picture, we've aligned every intervention with our goal of boosting student outcomes across the board.

Throughout my time as a principal, this leadership approach has transformed how I lead, allowing me to navigate with precision, purpose, and alignment with our systemic goals. By handling these layers, we've propelled our school toward a path of excellence, nurturing a culture of continuous improvement and cooperation. My experiences have really shown me the need to cultivate a supportive and inclusive environment. One where every staff member is empowered to make a lasting impact on their students' lives.

—Dr. Bruce A. Tinor, BCBA, LBS
School Administrative Leader, Special Education Principal

Preface

If you want to teach people a new way of thinking, don't bother trying to teach them. Instead, give them a tool, the use of which will lead to new ways of thinking.

—COMMONLY ATTRIBUTED TO R. BUCKMINSTER FULLER

In the spirit of R. Buckminster Fuller's wisdom—that to truly change someone's thinking, it's more effective to hand them a tool than lecture them on new perspectives—we present the second edition of our work. This edition is not merely a continuation but an evolution, born from our combined decades of experience and dedication to advancing the field of educational and Organizational Behavior Management (OBM). Herein, we aim to offer not just insights but practical methodologies to those on the front lines of education: instructional leaders, coaches, and all professionals committed to nurturing teacher success.

Our journey, which crosses the diverse terrains of public and private education sectors, business leadership, and even professional sports, has confirmed one steadfast belief: The process we advocate is universally applicable and profoundly effective. From assisting Fortune 500 CEOs in achieving both fiscal growth and cultural transformation to supporting school leaders in revitalizing underperforming institutions, the breadth of our process's applicability is vast. Its efficacy extends to enhancing communication for children with disabilities, elevating instructional performance among teachers, and even aiding professional athletes in securing championships. The question you might ask is, "What is the linchpin of such a versatile and potent process?" The answer lies in the foundational science of human behavior, known as Applied Behavior Analysis (ABA).

Our commitment is to bridge the gap between the rich—albeit sometimes intimidating—complexity of ABA and its practical application in the educational landscape. We strive to demystify the science, making it accessible and actionable for educators and leaders alike. The aim is to transform how instructional leaders and coaches engage with educators to equip them to optimize every interaction to foster growth and excellence.

In this edition, we have taken deliberate steps to enhance the practical utility of our work. Recognizing the value of concrete examples in illustrating complex concepts, we have significantly expanded our use of case studies and real-world scenarios.

These additions are designed to clarify the principles of ABA and also to demonstrate their application across a variety of contexts. The addition of a dedicated chapter on creating your own coaching system represents a direct response to the feedback and needs of our readers. This chapter provides a structured approach to designing and implementing effective, science-based coaching strategies tailored to specific organizational and educational environments.

Our objective with this edition is twofold: to deepen the reader's understanding of the foundational principles of behavior science and to offer a clear, actionable framework for applying these principles in a manner that is both effective and sustainable. The Deliberate Coaching model, which lies at the heart of our approach, is presented in greater detail, providing readers with a comprehensive guide to developing and refining their coaching practices.

In essence, this work is a toolkit designed to empower educators and leaders with ABA-based strategies for developing a nurturing, effective, and high-performing educational environment. It serves as a roadmap for applying the principles of behavior science to foster not only improved teacher performance but also a culture of support, engagement, and positive outcomes within schools.

Structured to cater to a wide audience, from K–12 instructional leaders and coaches to stakeholders at all levels of the educational system, the book is divided into three key sections. Each segment builds upon the last, beginning with an examination of the current state of education, moving through the foundational principles of behavior science, and culminating in the presentation of our Deliberate Coaching model. This model is the heart of our approach, offering a concrete framework for applying behavioral science in a way that is both systematic and empathetic.

Our narrative addresses the urgent need for a shift from reactive to proactive leadership within schools, advocating for a constant stream of support and recognition for educators that extends beyond periodic evaluations. This proactive stance, we argue, is fundamental to developing a culture where educators feel continuously valued and supported—not just in response to challenges or shortcomings.

This book is an invitation to all involved in the educational ecosystem to adopt a coaching mindset, one that can originate from any role within a school's structure. By integrating the science-based strategies detailed within these pages, readers are equipped to make a tangible and lasting impact on the performance and well-being of both teachers and students, fostering an educational environment characterized by support, engagement, and success.

In closing, our hope is that this second edition deepens your understanding of the practical applications of ABA in education and also that it inspires a commitment to fostering a positive, scientifically informed school culture. Here's to embarking on a journey together toward a more supportive, effective, and enlightened approach to educational leadership and coaching.

Acknowledgments

In refining this second edition, our journey has been significantly influenced by the broader educational and behavior-analytic community. Their collective wisdom has deeply enriched our roles as Organizational Behavior Management practitioners, offering us a treasure trove of insights that have directly and indirectly shaped our understanding and approach.

Our sincere appreciation goes to KeyPress Publishing, and in particular, Mary Sproles Martin, Ashley Johnson, and Dr. Darnell Lattal. Their detailed review and invaluable feedback on our manuscript were instrumental in elevating the quality and accuracy of our work. Their expertise not only enhanced our content but also guided us through the nuanced process of refinement.

Following this, we wish to extend our individual thanks to those who have personally supported us throughout this endeavor. As two authors united in purpose but inspired by different journeys, we recognize the unique contributions of several special individuals who have impacted our work and lives in profound ways.

Paulie

I have been fortunate to be surrounded by and learn from some amazing people across the fields of education, human services, and combat sports. Whether directly or indirectly, the following folks have either provided me an important opportunity or furthered my knowledge and ability as a leader, coach, or as a person. They include: Dr. Alex Edmonds, Brad "One Punch" Pickett, Bruce Gorman, Chris Frick and family, Din Thomas, Francisco Gomez, Frank Krukauskas, John Taylor, Dr. Jose Martinez-Diaz, Dr. Ken Wagner, Kenny "Deuce" Garner, Joel Garcia, Dr. Kisha Bellande-Francis, Leigh Martin, Luigi "The Italian Tank" Fioravanti, Mario "Big Hurt" Rinaldi, Mike Brown, Milton Lacroix, Dr. Nic Weatherly, Renato Tavares, Roger Krahl, Dr. Scott Neil, Dr. Tom Gollery, and my business partners in The Behavioral Toolbox, Anika Costa and Matt Cicoria.

To my biological father, Paul Gavoni, whose journey through life was a complex tapestry woven with the threads of mental health challenges. In 2023, he found peace in a way that left us with heavy hearts and a deeper understanding of the struggles many silently bear. Though he did not raise me in the traditional sense, his unique way of offering support and expressing love left an indelible mark on my heart.

Similarly, my grandparents, Frank and Emma Gavoni, who departed this world many years ago, continue to occupy a cherished space in my memory. Their legacy is one of enduring warmth and unconditional love, serving as a beacon of guidance and affection in my life's journey. Even in their absence, the values they lived by—compassion, perseverance, and kindness—remain a guiding light, shaping the person I strive to be.

To my parents who raised me well, John and Susan Parry, the depth of my gratitude is immeasurable. From my earliest memories, you instilled in me the profound importance of lending a hand to those in need, a lesson that has become the cornerstone of my personal and professional life. Your teachings have shaped my worldview, guiding me to pursue a path where making a positive difference is my highest priority. Beyond the invaluable lessons, your unconditional love has been my constant source of strength and reassurance. In moments of doubt, your belief in me has been unwavering, providing the foundation upon which I've built my ambitions and achievements. Your support and love have been the greatest gifts, inspiring me to extend the same compassion and assistance to others that you have so lovingly gave to me.

And most of all, my heartfelt thanks go to my wife, Nikki Gavoni, and my son, Niko Gavoni. Your constant support has been the bedrock of this marvelous journey. Nikki, your unwavering faith in my work and your endless encouragement have been the light guiding me through the challenges and triumphs alike. Your partnership is a wellspring of inspiration and love that fuels my dedication and passion. Niko, your presence has brought an immeasurable joy and perspective into my life, reminding me of the importance of the work I do and the legacy I hope to leave behind. Together, you both make every step of this journey deeply fulfilling. Your love and support have made all the difference, turning challenges into opportunities and dreams into realities.

Nic

Any joy and success I experience in my personal and professional lives are because of the support and guidance I get from my lovely wife, Allison, and the inspiration and love I receive from my brilliant and beautiful daughter, Emma. We work as a team and I wouldn't, and couldn't, have it any other way. My wife brings out the best in our family and our daughter is the best of us.

I would like to thank the faculty, staff, and students of Western Michigan University with whom I worked while studying behavior analysis for the first time and the team of faculty, staff, and students at Endicott College who continue to research and apply the type of valuable behavioral tools needed to impact the world. I'd also like to acknowledge the students at the Florida Institute of Technology who helped with the first edition of *Deliberate Coaching*, particularly Jacqueline Noto for her review of manuscript citations and coaching literature, Joshua Addington and Nicole

Adriaenssens for their assistance gathering references, and Chad Kinney, Dennis Uriarte, Nelmar Cruz, Cledia Caberlon, Kayce Nagel, Taylor Hetherington, Noell Jankowski, and Estefanía Alarcón Moya for their assistance analyzing and refining coaching concepts.

I would like to thank Dr. Paulie Gavoni for helping me continue to expand my appreciation for the science behind coaching and the application of these principles in a way that has been a great value to my research and practice and, in turn, something I hope will continue to benefit those I serve over the years.

Lastly, I would like to thank my grandmother, LaVonne, and my father, Daniel, both of whom left the world far too soon. Their toughness taught me to keep moving forward regardless of what life throws at me, their virtue taught me to always stick to my center, and their genuine nature taught me to never be afraid to stand for what I believe. I'm very proud to have been raised by them and dedicate my work to them.

SECTION 1:

The Current State of Education

1

Systems and Systematic Approaches

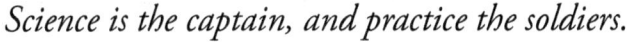

Science is the captain, and practice the soldiers.

—Leonardo da Vinci,
The Notebooks of Leonardo da Vinci

Each of us operates in numerous systems each day. Our homes are systems containing different people, requiring different duties (fun and chores and more), and presenting different outcomes. Our communities are made up of systems: different groups of people with different resources working together to achieve a common objective. And every school is a system. Schools enroll students into their systems to work with the teachers, staff, administrators, curricula, and resources. It is the job of the people and processes within your school system to produce students with strong learning repertoires that lead to successful achievement of standards. Each person in the system is valuable and linked in some way to this overall outcome of student success. We all work together to meet our goals.

Consider the educational system as a macrosystem, a vast umbrella encompassing numerous smaller systems such as organizations, districts, and schools. This complex structure integrates hundreds of processes, composed of thousands of tasks, all driven by myriad behaviors that must occur daily. The ultimate goal? Student achievement.

Complex, isn't it? This complexity arises from the interdependence of behaviors across various levels. District leaders support school leaders, who in turn support teachers, and teachers nurture students. In this intricate network, the success of students hinges on the seamless interaction of behaviors across diverse roles. This concept of interconnected behaviors aligns with what's known in behavioral literature as metacontingencies (Malott, 2003; Glenn, 1991), where each behavior is interconnected and reliant on another.

So, why this emphasis on interlocking behaviors and metacontingencies? Picture it akin to a car assembly process, governed by a series of "if-thens." Each individual on the assembly line has a specific task, and the completion of each task sets the stage for the next. Ultimately, the culmination of these tasks results in a vehicle ready for the market.

Similarly, the educational system operates with a myriad of professionals—from politicians to provosts, deans, and professors in higher education, and on to district leaders, school leaders, guidance counselors, and teachers. All are united in the singular aim of improving student achievement. These processes play out across federal, state, and local levels. Ellis and Magee (2007) delineate key hierarchical units within the educational system:

- federal agencies crafting laws to ensure the implementation of state directives based on federal guidelines
- state education agencies designing educational programs
- school districts setting qualification criteria for hiring teachers
- schools, in site-based management systems, engaging in the hiring of teachers
- classroom teachers executing the prescribed educational programs

According to the U.S. Department of Education's 2023 mission statement, the aggregate product of all these interconnected units "is to promote student achievement and preparation for global competitiveness by fostering educational excellence and ensuring equal access" (U.S. Department of Education, 2023b). The Organisation for Economic Cooperation and Development (OECD) classifies the assessment of educational systems into three categories (OECD, 2020):

- Output Indicators: focusing on student achievement measures
- Attainment Indicators: examining aspects like graduation rates
- Impact Indicators: assessing the employment status of graduates in certain job levels

Additionally, leading indicators, akin to mile markers guiding the way, are crucial. These include monitoring financial and material resources, and tracking daily attendance, tardiness, and student performance scores, all pivotal in informing progress and directing interventions aimed at enhancing student achievement.

What and How Students Are Taught

When we used the assembly line analogy to illustrate the interconnectedness of various processes, tasks, and behaviors, it wasn't too far from the truth. Many believe the current model of education in the United States continues to reflect the training mentality where students are taught the essential skills required to become a successful factory worker. In the factory model, students are grouped together based on their age, not their abilities. Sound familiar? During the early 20th century, factory-working skills were required to be competitive in a global market for the majority of folks who were less affluent. But times have changed dramatically, and the education system has been too slow to catch up. Unfortunately, the longer a system is in place, the harder it is to change.

In 1984, B. F. Skinner published an article titled "The Shame of American Education." In it, he noted that technology capable of doubling the rate of learning in schools across the country had been available for 20 years. "Most current problems could be solved if students learned twice as much in the same time and with the same effort" (p. 947). There is a scientific pathway to learning that every teacher can benefit from to motivate students, support progress, and maximize learning opportunities to accelerate learning and promote the retention of concepts taught. This scientific foundation and support for teachers should be a primary focus of school reform.

Now here we are, more than 55 years since that technology was made available, and we are in a similar boat. But instead of embedding systems that help schools become better places to teach and learn critical standards, we've simply increased the number of standards being taught, reduced funding for the arts, disregarded the fact that individuals learn at different rates by creating a timeline of expected learning, and developed numerous punitive sanctions to enforce ill-advised policy. This scattershot approach to standards-driven instruction has had a negative impact on the development of fundamental knowledge and skills to a level of mastery (accuracy of responding) and fluency (speed and accuracy).

And fluency is a prerequisite for the type of accelerated performance Skinner was talking about when discussing the need to save time and effort in the learning process. Much like an arrow shot by an archer, if fundamental skills are even slightly off in the beginning stages of learning, learners will fall far short of hitting the bull's-eye.

The effectiveness of accelerating performance by systematically developing fundamentals was illustrated well in a study on teaching elementary students to read (Alessi, 1987). The investigator found that by blending (sounding out letters to make words) only 40 or so letter-sound combinations in the English language, a child would be able to read almost 500,000 words. Can you imagine the time teachers could save by teaching the 40 sounds instead of 500,000 words?

This concept is rooted in the phenomenon known as adduction (spontaneous learning), whereby complex learning can emerge without explicit instruction when the right fundamentals are combined, sequenced, and taught to fluency. The systematic development of fluency is critical as it fosters the capability of individuals to operate in a skilled and productive manner within their usual environments (Binder, 1996).

Don't take our word for it. Project Follow Through, one of the most extensive educational experiments ever performed by the federal government, compared 22 different models of instruction across more than 200,000 children and 178 communities composed of students from diverse demographic, ethnic, and socioeconomic backgrounds (Engelmann, 2007). The results were clear: Direct instruction, a research-based approach to instructional design and implementation rooted in behavior analysis (Binder & Watkins, 1990), had a significantly higher impact on academic achievement than any of the other programs. Students participating in the research also demonstrated higher self-esteem and self-confidence than students who received different models of instruction. Follow-up research found that students who received direct instruction:

- continued to outperform their peers
- were more likely to finish high school
- were more likely to pursue higher education

Oh, and by the way, this research was originally conducted between 1968 and 1979. Yet here we sit, nearly 50 years after the last of these amazing outcomes was established, and direct instruction, what one professional called the "dirty little secret from the biggest education study ever" (Lindsay, 2014), is only being used in a handful of schools.

The question you are likely asking is, "Why, oh why, oh why aren't major changes within the educational system happening?" Unfortunately, the answers are deep, varied, and beyond the scope of this book. However, we will take a brief shot at one or two potential reasons below, understanding the reasons might vary across each educational system, and then further reflect on them in future chapters.

The first reason major changes are not happening has to do with systematic change management. While studies such as Project Follow Through yield amazing data, research, by its nature, is conducted by researchers, not change agents. Approaches to getting people to embrace and engage in change often involve a long and sometimes challenging process that has been documented in volumes of literature by a plethora of authors and practitioners. In John P. Kotter's popular book *Leading Change*, first published in 1996, Kotter highlighted eight errors in organizational change:

- allowing too much complacency
- failure to create a sufficiently powerful guiding coalition
- underestimating the power of vision
- under-communicating the vision
- permitting obstacles to block the new vision
- failing to create short-term wins
- declaring victory too soon
- neglecting to anchor changes firmly in the corporate culture

It's easy to imagine that all or most of these errors might be partially responsible for the absence of direct instruction in most schools across our nation.

The second reason has to do with habit. You've probably heard the saying "You can't teach an old dog new tricks." Well, one might consider the educational system the old dog and direct instruction the new trick. The longer a system has been in place, the harder it is to change—that is, if you aren't familiar with the science of human behavior as applied to organizations. We'll talk more about the underlying principles of behavior in future chapters. For now, suffice it to say that there was clearly little to no science behind the delivery of direct instruction to the masses in the past and, thus, little to no support to transfer these educational standards into practice. Had there been, we likely wouldn't have written this book.

We can do better. Instead of trying to enforce policy through threat of punishment, where behavior decreases because of something undesired following the behavior, we must seek to systematically transfer the most effective practices for learning. We must support those using these science-based educational practices and set them up for success. The technology for accelerating the acquisition of knowledge and skills exists, and it should be taught to fluency in higher education so that it can be effectively transferred into the classroom.

Case Example: The Systematic Application of Behavior Analysis

Promises are a dime a dozen in education, but guarantees are rare. It's hard to imagine a school that guarantees student growth, let alone one that is willing to put money on the line. Take a look at the average Morningside Academy student grade level growth in reading comprehension during the 2015–2016 school year—jaw-dropping to many (see Figure 1.1).

What makes this possible? The answer is a comprehensive learning model based on mountains of research on what works in education. In 1980, Dr. Kent Johnson founded Morningside Academy, a behaviorally based laboratory school that helps elementary and middle school students catch up and get ahead. Inspired by John Dewey's laboratory schools, Morningside serves Tier II general education and mild special education students—the forgotten 40% of students in the United States. These kids present with a broad range of learning challenges and far too often fall through the cracks of the American education system.

The success of children at Morningside is the outcome of years of research and development into the Morningside Model of Generative Instruction (MMGI), a generic learning model that has produced profound results for a wide range of learners. MMGI is based on the evidence-based best practices that have proven effective with learners at the laboratory school and at schools across North America.

MMGI is built on five pillars: assessment, curriculum, instruction, precision teaching, and generative responding (Kieta, 2020). First, a research-based curricu-

Grade Level Growth in Reading Comprehension for Students Two or More Years Behind, 2015–2016 School Year

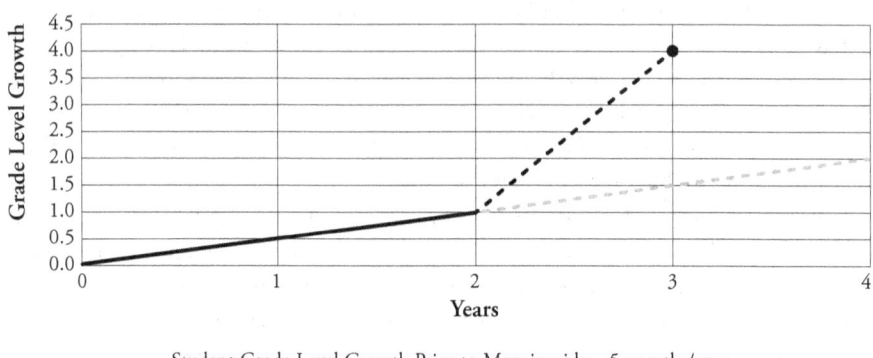

Figure 1.1. Average student grade level growth per year in the area of reading comprehension for students who were two or more years below their grade level prior to entering Morningside Academy; their projected growth without the Morningside intervention; and their actual growth at the end of the year. These data reflect growth on the Iowa Tests of Basic Skills for the 2015–2016 school year. Adapted from Kieta (2020).

lum is selected that focuses on specific component skills (assessment). Morningside instructional designers examine the types of "holistic" and "real-world" activities that often take place in today's general education classrooms. Then, they select and design instructional programs that will best teach the component skills that students need to be successful in those general education settings (curriculum).

Second, students are placed in classes by what they can and can't do, rather than by their age. This process of homogeneous achievement grouping allows for optimal instructional groups and facilitates the rapid growth that students at Morningside make (instruction). After selecting the best curriculum and placing students in the best learning environments, Morningside teachers directly and explicitly teach components using *mathetics*, the underlying science of direct instruction. Teachers use this science to ensure that students can accurately complete a targeted objective before moving on to new objectives.

The fourth pillar is precision teaching. Once students are accurate in their performance of a newly taught skill or repertoire, they engage in systematic, daily practice in order to build fluency. At Morningside, mastery does not just mean high accuracy, but high accuracy plus high frequency—or speed. Mountains of research show that completing skills in a timely fashion is a key indicator of retention, generalization, and more.

Lastly, Morningside systematically creates learning environments that produce *generativity*—new learning without instruction (generative responding). The curriculum is arranged in such a way that students can be given new challenges that they will be successful in solving. This means putting them into structured versions of the "holistic" and "real-world" activities that are prominent in general education. The big difference is that Morningside students have shown that they have all of the skills necessary to be successful in environments that can often be overwhelming. One way this is done is through teaching students problem-solving strategies, such as Talk Aloud Problem Solving, a problem-solving and reasoning curriculum developed by Morningside principal and associate director Joanne Robbins.

The desired outcome of all of this work is generativity. By focusing on critical component skills, guaranteeing mastery via explicit and direct instruction, using precision teaching to ensure fluency, and systematically arranging learning environments, MMGI consistently puts students at Morningside into positions where they find they can do something they've never been directly taught.

A common challenge from Morningside teachers is "I bet you can figure that out on your own!" And nothing builds intrinsic motivation and drive better than doing just that. It's clear that students cannot possibly learn everything they'll ever need to know in the relatively short time they spend in school, so helping them construct repertoires that allow them to meet ever-changing demands is critical to their long-term success. Focusing on prerequisite component skills—which are far too often overlooked—and teaching problem-solving and reasoning strategies yields agile, confident, persistent learners.

Labels do not and should not define children. Rather than trying to compensate for or work around learning challenges, MMGI uses a scientific approach to directly target and strengthen those areas. Forty-five years of data support this view, and Morningside shows no sign of slowing down. MMGI continues to be developed, helping the next generation of students build the repertoires they need to be successful learners and adults.

Behavior analysis provides a robust scientific foundation for understanding learning and performance. When someone learns, it's a direct result of applying these behavioral principles, whether we're conscious of it or not. By deepening our understanding of behavior analysis, we gain greater control over educational environments and enhance learning outcomes. It's crucial for the evolution of educational systems to incorporate science-based teaching methods. Moreover, supporting educators in adopting these practices through systematic, habit-building coaching is a game changer. This book lays out the blueprint for such an evolution, promising a future where education is not just about imparting knowledge but about optimizing the very process of learning itself.

2

State of Education

The man who moves a mountain begins by carrying away a small stone.

—COMMONLY ATTRIBUTED TO CONFUCIUS

Behavior analysis can offer a wealth of evidence-based tools that have proven to be effective at improving a range of employee behaviors in countless settings. But these tools will only work if they can be integrated correctly into your particular school. Your school has goals specific to your school, a staff infrastructure that works for your school, and a long history of what's been working and what hasn't been working. Evaluating the state of education over time will leave you to consider a number of factors and potential barriers that can stand in the way of long-term improvements.

Every school has a certain number of people responsible for the performance of teachers, and the number of these individuals will have an impact on how much time is available to shape and maintain teacher performance. It's not the job of school leadership to simply *evaluate* performance; school leaders are in a position to help *build* the type of teacher performance that will lead to the outcomes and culture necessary to hit your goals. But whether they assign a principal, vice principal, counselor, or even a district coach for added support, there are a number of things that need to be done—and only so many hours in the day. If you can't find a way to overcome these constraints, then leadership strategies might be left to trainings, evaluations, and putting out fires instead of proactively building and supporting teacher performance.

Time is money, and funding is always an issue when it comes to finding staff for coaching and resources for training and support systems. Whether you're making the case for funding at the district level or trying to manage the funds at the school level, this is a barrier that can get in the way of any leadership engagement. The solution isn't to ignore consistent support in favor of quick training options; finding efficient ways to integrate your coaching system is a critical initial step.

In the current state of education, policymakers have focused on desired outcomes and not as much on how to get those outcomes. This is a slippery slope that leads to ignoring critical life and academic skills and instead fixates on, for example, answers on standardized tests.

Tale of Two Classrooms

Mrs. Kohn has been a teacher for 25 years and has an approach that seems to work for her. Students aren't very active during her class, but they maintain adequate test scores despite a few instances of minor student disruptions each week. While the test scores aren't exactly the highest in the district, they're enough to get by. Students who perform at a high level are left alone. Students who act out or perform at a lower level get more attention.

Mr. Miller's classroom runs a bit differently. He's been there a few years less, but he's managed to gain quite a reputation for turning out productive students who speak highly about his methods. He monitors progress and pays attention to how many opportunities each student has to engage in activities throughout the day. Some students like certain activities more than others, and he uses information on their preferences to keep them motivated. His students say that learning is fun—good things happen when they're learning. They're never bored. They seem to consistently go above and beyond minimum expectations.

So what's the key difference in what they're doing? Mr. Miller focuses on behaviors, not just outcomes. He focuses on engagement—ensuring that each student gets ample opportunities to succeed.

It's not the job of school leadership to simply evaluate performance; school leaders are in a position to help build the type of teacher performance that will lead to the outcomes and culture necessary to hit your goals.

High-performing students shouldn't be ignored because they seem fine on their own. All students deserve opportunities to grow, all students have unique needs, and all students have their own likes and dislikes. You need high-achieving students, but it's the way you get there that offers the critical opportunity for growth.

The Blame Game

While some schools can tout high achievement, too many struggle with making desired achievement gains. When this happens, the blame game begins. Policymakers blame and pressure state leadership; state leadership blames and pressures districts; district leadership blames and pressures school leadership; and school leadership blames and pressures teachers. As this happens, students fail to reach their potential as frustrated parents blame the teachers; frustrated teachers begin pointing their fingers back at the parents; and leadership on every level receives the finger until it bounces back to policymakers. Policymakers under pressure decide they need to do something, so they draft a new policy and begin the whole blame game again.

The blame game is real. And it's painful for all involved. Let's take a closer look at an all-too-common scenario in failing schools.

> After graduating from college, a group of first-year teachers is hired by a hopeful principal of a Title I school. The principal is assigned to the school after the former principal is fired by the district because the school failed to make adequate gains 3 years in a row. The principal wants seasoned and culturally diverse teachers, but high-poverty schools in her area have a bad reputation, so her selection pool is thin. The group of teachers begins the school year aspiring to help students in poverty achieve.
>
> While the first week isn't so bad, seemingly small issues begin to wear on the staff. In the second week, most begin struggling. Lesson planning, grading, managing student behavior, and instructing based on the district's evaluation model quickly erode their morale. Their high hopes and aspirations are replaced with stress that progressively evolves into panic as the principal walks through the classrooms, pressuring teachers to boost student achievement. Attempting to help, the principal leaves each teacher with a to-do list for improving instructional practices.
>
> The new teachers make it through the first quarter. While they have been assigned mentors, most of the mentors are too busy struggling in their own classrooms to provide much guidance. Those who do receive guidance and try some of the suggested techniques aren't getting the desired outcomes. Many in the group are now questioning their career choices. What seemed to be the first steps toward a life's dream have rapidly disintegrated into a nightmare.
>
> While they put their tough faces on, many cry at home and are anxious about coming into school each morning. A few actually break down crying to the principal. The principal, while sympathetic to their plight,

is receiving flak from the district as achievement scores aren't where they should be. Asking for support, the district provides the school with two instructional coaches who each serve five other schools. One of the coaches is a reading coach; the other is a math coach who has some background in classroom management.

After meeting with the principal to identify the group of teachers in need, the coaches observe and complete individual checklists to rank 50 instructional and classroom-management practices as "Not in Place," "Somewhat in Place," and "In Place." They then debrief with one another to determine the best approach for helping the targeted teachers. Some of the teachers scored an "In Place" on a couple of the targeted elements; however, most of the elements are "Somewhat in Place" or "Not in Place." The reading coach thinks the problem is classroom management, while the math coach blames most of the issues on instructional strategies related to reading.

Failing to reach a compromise on what high-impact instructional behaviors should be targeted, they use a scattershot approach, identifying 35 out of the 50 elements that require improvement. They hand the checklist to the principal and then inform her they won't be back until next week.

The end of the year rolls around. The principal is tired, and many of the teachers say they won't be returning. District leadership, unhappy with student achievement and teacher attrition, decides the principal isn't a good match for the school and hires another one in her place.

The new principal, enthusiastic and optimistic, seeks to hire seasoned and culturally diverse teachers; however, she struggles because high-poverty schools in her area have a bad reputation, so her selection pool is thin. She ends up hiring a group of first-year teachers to fill her remaining positions. The new teachers begin the school year aspiring to help students in poverty achieve … and the blame game, like Groundhog Day, begins again.

This story is like a bad movie, except that students and educators can't walk out of the theater and ask for a refund. Too many are stuck. If you've not experienced this as an educator or even a student, consider yourself fortunate. Many educators in struggling schools feel as if they're rowing upstream. They soon fatigue and leave, at the expense of themselves and our children. President Obama echoed the sentiments of many when he stated, "A child's course in life should be determined not by the zip code she's born in, but by the strength of her work ethic and the scope of her dreams" (Slack & Oken, 2014).

A personal reflection on the state of education in your region, state, or nation will most likely lead you to find opportunities for improvement. There are many ways we can improve education, some of which we hit on in this book and some that extend far beyond it. While it's important to vote, get involved with the school board, and find other ways to make changes within various levels of your school system, this book will hopefully show you how to provide better support for the valuable personnel who make up the school cultures to which we send our children.

The Fight for Knowledge

Cus D'Amato, the famed boxing coach who trained multiple champions, including the infamous Mike Tyson, said of Tyson, "A boy comes to me with a spark of interest, I feed the spark and it becomes a flame. I feed the flame and it becomes a fire. I feed the fire and it becomes a roaring blaze" (Heller, 1995, p. 63). This "spark" is fundamental to getting somebody to engage in something. But when people are thrown into deep waters unprepared, their spark is often dowsed. As we've already highlighted, this occurs far too often in education as leaders and teachers are thrown into deep waters, unprepared to meet the demands of the school and classroom.

In the beginning stages of development, the original spark might be enough to get educators moving; however, it is not enough for sustainability. If educators are to succeed, trainers must take great care to purposefully (with intentional design) and systematically develop skills required for them to reach established goals. If the educator is unable to observe meaningful consequences that result from each skill, the skill and the motivation to use it are likely to fade quickly. To ensure that people will accomplish targeted goals and achieve desired outcomes, focus must be placed on what skill sets should be developed and how the individual skills will be developed.

BEHAVIOR = BELIEVING

Valuing goals is part of the formula for successful performance. As such, it is extremely important to involve people in their development whenever possible to increase engagement in schools. However, for the formula to be effective, it is equally as important that educators believe they can perform a given task, believe performing this task will help them achieve the desired results, and believe professional developers and coaches will help them improve their performance.

This belief in their ability to accomplish a particular task is what Albert Bandura (1997) termed "self-efficacy." The concept continues to receive attention (e.g., Klassen & Chiu, 2010; Pfitzner-Eden, 2016a, 2016b; Tschannen-Moran & Hoy, 2007) as a predictor of teacher success. But belief, regardless of who you are or what you do, is not something that is built by simply telling people they can do something. Belief occurs as a result of behavior. And while believing and having confidence might get somebody to try something, they aren't sufficient for lasting success. It is behavior that achieves meaningful outcomes that ultimately leads to success and self-efficacy.

In other words, if teachers repeatedly behave in some way, they are repeatedly able to observe the desired outcome as a result of these mastery experiences. As a result, their belief in their ability is strengthened. And if this outcome is valued by the individual, this increases the likelihood that they will continue performing in order to achieve the result. In the end, the goal of training isn't about getting people to believe. The goal here is to develop skill sets to the point where if an individual performs a task, they will be successful. And though many struggling educators lack a sense of self-efficacy, or belief in their ability to achieve certain outcomes, because they've not yet had these mastery experiences, they all truly want to be successful. They fight every day for it!

FIGHTING FOR EDUCATION

Educators are very much like fighters: Instead of arming themselves with combative skills, they arm themselves with knowledge and instructional skills as they battle to empower their students with the repertoire they need to be successful personally and professionally. Telling them what to do or how to do something is insufficient for success (Gavoni, 2015).

Can you imagine a boxing coach throwing one of his novice fighters into the ring with a professional after simply telling her what to do and how to fight? Do you think this would result in the fighter strengthening her skill set or her subsequent belief in her ability that would result from successful experiences? Do you think the fighter would value the boxing coach? Nope. Rather than learning anything, the novice fighter would likely employ whatever skills exist in her current repertoire to avoid taking a beating. Afterward, the novice fighter would probably view the coach as callous. And the fighter might falsely believe she doesn't have what it takes to be "a contenda."

We can't understand why so many teachers and leaders are just thrown into the schools to take a "beating" like the novice fighter. All too often, universities and teacher-preparation or induction programs tell educators what and how to do what they're supposed to do with the expectation that this will lead them to successfully do it. This is counterproductive and costly on many levels to the teachers, staff, coaches,

"For the things we have to learn before we can do them, we learn by doing them. Knowledge is of no use unless you put it into practice."

—Aristotle, circa 350 B.C.E.

leaders, school, and—most of all—the students. Skills must be developed through regular practice.

For this reason, it is essential that those associated with training employ systematic strategies for developing skills (using organized plans or repeatable procedures) that will eventually allow educators to perform in a way that leads to a desired outcome. These strategies should focus on strengthening educators' performance to the point where they can have mastery experiences by systematically developing their declarative, procedural, and conditional knowledge. While the terms for these kinds of knowledge are increasingly popular in education, we like to break them down into their simplest forms: what to do, how to do it, and when to do it.

DECLARATIVE KNOWLEDGE

While *instructing* people on how to do something is insufficient for getting them to perform successfully, it's still an important part of the formula for developing successful performance. In education, understanding what to do is known as *declarative knowledge*. Declarative knowledge has been described by some as the factual information accumulated (Stürmer et al., 2013). This is the "know" associated with learning. And what's a simple way to determine if somebody knows something? Behavior! Specifically, *verbal* behavior! Can the person communicate the information to us verbally or in writing?

One simple way to assess if somebody "knows" what to do is to simply ask them, "What are you supposed to do?" If they can't describe each step, this is a good indicator that they may not know what to do, let alone be able to do it. And note that we're not simply asking "Have you been trained?" but instead asking to see the behaviors associated with the required task. Just because someone has been trained does not mean that the training worked.

When providing the "what," it's also helpful for educators to understand the "why," as this oftentimes directly impacts the "will." Understanding the rationale for a given behavior is part of setting that behavior up for success. Folks who are told to do something that doesn't make sense to them are far less likely to focus on learning how to do it during training or on performing it in school. In contrast, folks who are taught to understand the why in the beginning are far more likely to engage in the learning and skill-development process as they come to understand how the skills are linked to meaningful outcomes.

You want people to participate because they see value in participating, not simply because they're told to participate. That is how to build habits and work to ensure people are engaging in desired behaviors even when nobody is around—because they want to. Because it works for them.

A good coach who is seeking to develop a teacher's knowledge might say, "Here is why. Try it this way (*modeling*). What are your thoughts?" It's important to know what your educators are thinking. Their thoughts can act as a formative assessment to guide future development efforts. Their thoughts will also affect their willingness to participate in future talent development linked to procedural knowledge.

Good coaches frequently test their teachers' knowledge by asking them questions about the skills they're performing—questions such as, "Why should you ask a question related to a lesson *before* calling on a student as opposed to calling on a student first and *then* asking the question?" The answer in this case is to engage all students in thinking about the question. Calling on one student first and then asking can inadvertently make other students less engaged. Helping the teacher understand this increases the likelihood that they'll ask the question first. They don't want their students to disengage as this can sometimes increase misbehavior and will negatively impact achievement!

PROCEDURAL KNOWLEDGE

Knowledge acquisition is the core element of many university- and district-based teacher-preparation programs (Darling-Hammond, 2006); however, skill acquisition is too often overlooked. In the education world, knowing how to do something has become known as *procedural* knowledge. You might consider this the know-how that connects what educators have learned (declarative knowledge) to what they do.

Trainers might initially explain what a teacher should do and why. However, this consumes a very brief chunk of the time allotted to training. Good trainers develop training regimens and require the teachers they're supporting to build knowledge through practice linked to successful outcomes (e.g., increasing student engagement). By practice, we mean trainers systematically develop teachers' skills by requiring them to engage in the repetitive rehearsal of those skills. If we wish for any educator to perform successfully within the school, they must have the opportunity to practice critical skills to a standard of proficiency.

However, as Vince Lombardi is often quoted as saying, "Practice does not make perfect. Only perfect practice makes perfect." This statement is pretty accurate. In order to perfect a skill, performers must receive feedback. If a performer practices a skill without feedback, they might quickly become very good at a skill that is very bad!

Just as coaches in sports help athletes reach performance goals by creating conditions whereby the athletes become experts in their respective arts, teacher-preparation, induction programs, and training programs should have a goal of allowing for purposeful and precise practice. And when we talk about precision, we mean providing teachers with exactly what they need to be successful in the classroom through targeted training and specificity of feedback.

In Chapter 6, we will cover the topic of deliberate practice when teaching skills to children in the classroom. In sports, deliberate practice is the purposeful approach that distinguishes experts from the average Joe. The foundation of this practice is breaking down specific skills and training into chunks and allowing the performer to become fluent through repetition *and feedback*. Good athletes spend hours deliberately drilling the skills of their sport while coaches provide them with a steady stream of feedback about specific skills to shape their performance. Just like these athletes and students, educators *must* engage in rehearsal of a skill *with feedback* if they're going to develop skills to any level of proficiency. Unfortunately, this isn't happening enough.

Clearly one-shot "sit and gets" for educators are not enough. Though trainers in education don't necessarily need to understand the science behind talent development, good trainers understand that talent is progressively developed through multiple learning opportunities that include practice and feedback. But on top of not providing enough practice opportunities, many educational talent development programs often expect their talent to learn too much, too soon.

3

State of Teaching and Learning

Change the way you look at things, and the things you look at change.

—Wayne Dyer, blog post

Where We Are

Teachers have the power to make a substantial impact on the academic growth, health, and well-being of their students. But this is a tough job, and it comes with a lot of responsibility and stressors. Some studies rank teaching among the most stressful occupations (Johnson, Cooper et al., 2005). Not only does this reduce morale, attendance, and effectiveness in the classroom (Hakanen et al., 2006), but one study published by the *American Journal of Alzheimer's Disease & Other Dementias* (Lippa, 2013) actually found that teachers are much more likely than other professionals to be diagnosed with progressive speech and language disorders.

The landscape of teacher turnover and attrition has evolved significantly in the wake of the COVID-19 pandemic, revealing a complex web of factors that influence educators' decisions to remain in or leave the profession. Recent studies shed light on the heightened challenges faced by teachers during and after the pandemic, contributing to increased turnover rates and altering commitments to the profession.

In fact, a survey conducted by the National Education Association in January 2022 has unveiled a concerning trend among educators nationwide: A staggering 55% were contemplating an early exit from their profession. This alarming statistic marked a significant uptick from the 37% reported just a few months earlier in August 2021, signaling a widespread sense of disillusionment that transcends age, tenure, and roles within the educational ecosystem, from teaching to transportation to nutrition services for students. Notably, the survey also sheds light on an even more distressing dimension of this trend—the disproportionate inclination to leave among educators of color, with 62% of Black and 59% of Hispanic/Latino educators considering departure (Walker, 2022). This is particularly troubling given their already underrepresented status in the profession, highlighting the urgent need for systemic interventions to address the unique challenges and disparities that contribute to this disconcerting desire to exit the educational field.

And they aren't just threatening to leave. Research from the Brookings Institution using statewide administrative data has shown that teacher turnover climbed in 2021 and 2022, with concerns growing over the sustainability of the teacher workforce, especially in high-poverty schools and districts (Harbatkin & Nguyen, 2023). This research highlighted that about 30% of teachers who reported plans to leave their school did so the next year, indicating a significant predictor of actual turnover behavior (Zamarro et al., 2021).

In fact, *Chalkbeat*'s analysis of teacher turnover from eight states revealed that turnover in the 2021–2022 school year reached its highest point in at least 5 years, typically around 2 percentage points greater than before the pandemic. This variation in turnover rates among states underscores the widespread nature of this issue, with Louisiana and North Carolina experiencing notably high rates of teacher exits (Barnum, 2023).

The acute need for better data systems to understand and address teacher turnover is underscored by research published in *Education Finance and Policy*. This study suggests that while teacher turnover remained relatively stable through 2021, the economic insecurity caused by the pandemic may have influenced this stability. However, the possibility of increased turnover rates persists as challenges within the education sector continue (Bleiberg & Kraft, 2023). A more recent study of attrition rates and factors in Oregon private Christian schools echoed similar concerns (Davidson, 2023).

A RAND survey found that nearly one-quarter of teachers indicated a desire to leave their jobs at the end of the 2020–2021 school year, compared with an average national turnover rate of 16% pre-pandemic. This shift in teachers' commitment to remaining in the classroom reflects the profound impact of the pandemic on educators' professional lives and well-being (Harbatkin & Nguyen, 2023).

One study titled "Teacher Attrition: Differences in Stakeholder Perceptions of Teacher Work Conditions" examines the diverse perspectives of principals, teachers, and parents regarding workplace conditions and how these perceptions influence teacher attrition. It demonstrates the significant impact of differences in opinions on teacher involvement in decision-making, as well as resource and discipline management. These contrasting views within school communities can critically influence teachers' choices to either continue in or leave their positions, illustrating the complexity of factors affecting teacher retention. This study highlights the complexities of addressing these systemic issues to mitigate teacher turnover effectively (Harris et al., 2019).

Furthermore, a significant study published in *Frontiers in Psychology* investigates the psychological effects of the pandemic on teachers, revealing a strong link between increased stress, burnout, and the high turnover rates observed, particularly in high-need areas. The research utilized longitudinal data to demonstrate how the heightened stress levels during the pandemic have exacerbated teacher attrition. It advocates for a comprehensive strategy to bolster teachers' well-being, emphasizing that enhanced mental health support for educators could be critical in retaining them in the profession (Gillani et al., 2022).

And it's not just teachers. Principals are pivotal in spearheading school improvement efforts and fostering inclusive, high-achieving educational environments. Unfortunately, principal turnover has emerged as a significant concern within the educational sector, with implications that ripple across school communities. A study published in *Educational Administration Quarterly* (DeMatthews et al., 2022) found that the stability of school leadership is under threat, as about 20% of principals leave their positions annually, with even higher rates observed in low-income, racially diverse schools. This study delves into the impact of principal turnover on teacher attrition, particularly focusing on urban and high-poverty contexts and the phenomenon of repeated leadership changes.

The findings reveal a significant uptick in teacher turnover following principal exits, an effect that intensifies in high-poverty, urban schools; among less experienced teaching staff; and in instances of frequent principal turnover. These outcomes highlight the critical need for enhancing leadership stability as a strategy to curb teacher turnover.

Together, these studies present a compelling picture of the current state of turnover in education, emphasizing the need for comprehensive strategies to support educators and stabilize the teaching workforce. While the pandemic certainly exacerbated the situation, the stressors associated with being a teacher or principal have plagued the educational system for a long time. The pandemic only served to expose long-standing shortcomings. Addressing the root causes of turnover, including job dissatisfaction, stress, and burnout, is crucial for ensuring that all students have access to quality education.

How We Got Here

In our opinion, a big problem is that teacher and leader preparation and induction programs are primarily rooted in theory, with little practice and oftentimes zero coaching to help apply skills they learn to the school. We're not suggesting that theory isn't useful, but its utility is limited to the degree it can assist us in understanding the real world in real time. While theories might help us see the big picture, skills are required to link theory to the real-world, real-time applications. In education, this translates to student achievement. Check out the relatively well-known study by Joyce and Showers (2002) and Figure 3.1 regarding the generalization of skills into the classroom after a teacher is only provided theory.

What kind of classroom impact does simply providing theory and demonstrations to teachers have? None! None of the knowledge and skills transferred to practice! It's dismal, but it makes sense. If telling (theory) and showing (demonstration) were all it took for folks to learn, teacher success would be off the charts, and so would student achievement. Unfortunately, theory and demonstration are what too many teachers and leaders are getting. The revolving door of teacher and leader attrition bankrupts organizational memory at the expense of students and taxpayers.

It's no secret that high turnover undermines student achievement. We'll address this further in the next chapter, but when policymakers conceived the No Child Left Behind Act (NCLB) in 2002, teachers were left dangling in the wind as all eyes shifted toward the effectiveness of instruction in the classroom. High-stakes testing, adequate yearly progress, and 100% proficient demands established measures and goals without much consideration of the knowledge and skills of those required to produce these outcomes. Instead of focusing on strategies for developing and accelerating educator performance to improve student achievement, NCLB simply set high standards for students and then sought to hold schools "accountable" for establishing measurable success. More accurately, they blamed schools for not meeting unreasonable demands.

The assumption seems to have been that teachers weren't getting the job done because they were lazy or incompetent. So to light the proverbial fire under the butts of educators, "wise" policymakers with arguably little educational experience put pen to paper to right the wrongs of the naughty district, school, and classroom leaders. Districts were forced to double down on this by allocating a ton of money and resources toward improving evaluation processes to correct teacher performance.

The result might be best described as a chain reaction:

- Teachers were forced to teach to the test to avoid the ire of the school leader.
- School leaders were compelled to direct teachers' focus toward select "bubble" students who could help schools make the grade.
- District leaders leaned on school leaders to make the necessary gains.
- State leaders pressured district leaders to get the job done.
- High-paid consultants were brought in to guide leadership.
- Technology was purchased and scattered across districts.

Professional Development Elements

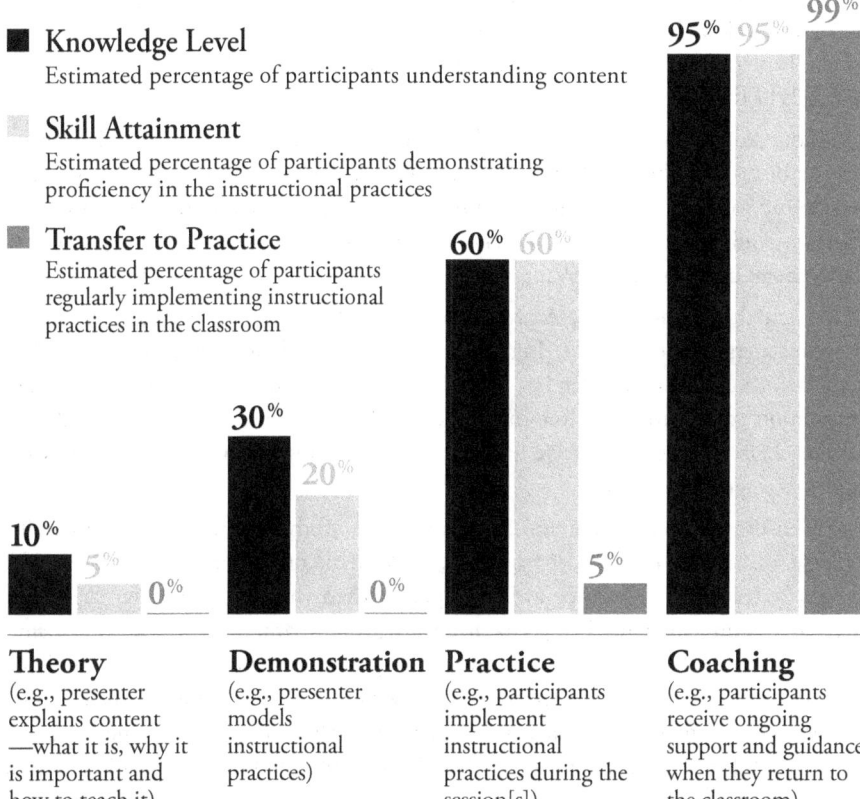

■ **Knowledge Level**
Estimated percentage of participants understanding content

Skill Attainment
Estimated percentage of participants demonstrating
proficiency in the instructional practices

■ **Transfer to Practice**
Estimated percentage of participants
regularly implementing instructional
practices in the classroom

Theory	Demonstration	Practice	Coaching
(e.g., presenter explains content —what it is, why it is important and how to teach it)	(e.g., presenter models instructional practices)	(e.g., participants implement instructional practices during the session[s])	(e.g., participants receive ongoing support and guidance when they return to the classroom)

Figure 3.1. *Why is coaching important? "Improving teachers' learning—and, in turn, their own practice and their students' learning—requires professional development that is closely and explicitly tied to [their] work. Coaching addresses that requirement" (Neufeld & Roper, 2003, p. 1). Figure adapted from Joyce and Showers (2002).*

The pressure was so great, it even led some good people to behave poorly, as evidenced by cheating scandals that rocked a variety of districts.

And after all is said and done, even with some promising developments on the horizon, people continue to blame teachers—pushing compliance by wielding accountability like a sword—instead of addressing a few key issues at the root. When students fail, we blame the teacher. When schools fail, we blame the leader. What about those who've failed teachers and leaders? Who is responsible for that? Let's stop blaming and start figuring out how to bring out the best in teachers so they can bring out the best in their students.

Preparation Programs

The reasons for lack of school progress and failure are wide-ranging and debated across industries. However, one issue that seems to get left by the wayside (while teachers and leaders are blamed) is how well teachers and leaders are being prepared for the demands of the classroom.

If our students are going to reach their maximum potential, we must seek to bring out the best in teachers and leaders. Teachers in the top 20% of performance have been found to generate 5 to 6 more months of learning each year than low-performing teachers. And as we noted earlier, school leaders have been found to affect student achievement by as much as 25%.

Given the impact of these professionals, it's shocking—to say the least—that many teachers and leaders entering the profession struggle to even pass certification exams. Teachers who transfer into education from another field without adequate preparation are two to three times more likely to leave the field (Ingersoll et al., 2014). And teachers who do not receive induction or mentoring are twice as likely to leave (Smith & Ingersoll, 2004).

Even those who are fortunate enough to receive formal education at a university struggle. According to a report by the U.S. Department of Education (n.d.-b), "far too many teachers report they are unprepared when they first enter the classroom after completing their teacher preparation program" (para. 3). This isn't surprising given that research suggests a clear lack of empirical evidence to support methods used in teacher-preparation programs (Walsh, 2006). Teaching is tough! Well-known educational researcher John Hattie (2009) hammers this notion home as he recognizes the flexibility and adaptability required in teaching methodologies rather than a strict, one-size-fits-all approach. Researchers investigating school leadership-preparation programs also cite a drastic need for improvements, as they have found a major disconnect between professor knowledge and the actual needs of the schools (Murphy & Vriesenga, 2006).

Teachers and leaders want and deserve better.

None of this is new. The issue has been recognized for decades. In preparation for the needs of the millennium, the American Council on Education (ACE) Presidents' Task Force on Teacher Education (2002) developed an action plan providing information on improving the quality of education offered to teachers and school leaders.

Even the data being collected by many preparation programs are practically worthless when it comes to measuring skill development by students. In research on school leadership preparation in the United States (Murphy & Vriesenga, 2006), researchers found that 80% of research on the topic focused on curricula or instructional strategies, which almost exclusively relied on student perception of:

- the degree to which program goals were met
- the extent to which program elements were valuable
- self-worth

And get this. According to the authors, in the nearly 2 decades since the 2006 publication of the article, leading journals in the field still have not published any research articles that directly evaluate the skills and knowledge acquired in preparation programs, nor are there any articles that measure changes in student performance as a result.

In fact, university-based principal preparation programs have been scrutinized regarding their effectiveness (e.g., Darling-Hammond, 2006; Murphy et al., 2008). Some researchers even suggest that the lack of leadership preparation programs is at the root of many issues across the nation's education system (Levine, 2005).

This is significant and echoed in The Wallace Foundation's report (Mitgang, 2008), which indicates that the school principal is the number one person for ensuring that quality instruction is provided across the school.

Now, if you paid attention to the references above, you will note a commonality— there are many references from older articles, suggesting the issue has been recognized for a long time; moreover, the bulk of the research was conducted pre-COVID-19. The COVID-19 pandemic has significantly exacerbated pre-existing teacher shortages across the United States, intensifying stress and burnout among educators forced to navigate the challenges of remote and hybrid learning environments (Schmitt & deCourcy, 2022). Enrollment in teacher preparation programs has seen fluctuations, with a noted decline preceding the pandemic and a complex landscape of stabilization or slight increases in certain areas post-COVID-19, reflecting a renewed interest in public service roles including education (Nguyen et al., 2022). Retention of experienced teachers has become a critical challenge, with increased turnover rates attributed to heightened stress, inadequate support, and often insufficient compensation, further complicated by the pandemic's demands (Schmitt & deCourcy, 2022).

The disparities in teacher shortages disproportionately affect low-income, rural, and urban schools, exacerbating educational inequities due to these institutions' tendencies to hire less experienced, uncertified, or out-of-field teachers (Podolsky et al., 2019). Unfortunately, the students who need the most are receiving the least.

And to echo what was addressed above: It's not just teachers—principal turnover in public schools has increased over the past 5 years, with seasoned leaders departing at higher rates compared to their less experienced counterparts. A comparison of public school principals' retention rates reveals a slight decrease, from 82% in the 2016–2017 school year to 80% in 2021–2022, indicating a growing trend of mobility and attrition within the profession (Taie & Lewis, 2023). This pattern is echoed in private schools, albeit with slightly higher retention. Furthermore, a notable proportion of principals expressed a willingness to leave the principalship for higher-paying opportunities, highlighting potential financial motivations behind the turnover.

Given the research and data above, coupled with our personal experiences, it's safe to say that many school leaders are in the same boat as teachers in that they're inadequately prepared to meet the leadership demands of the schools they serve.

Like a novice boxer who's thrown into the ring with a prizefighter to see if they've got what it takes, leaders and teachers are too often thrown into the school and classroom to sink or swim. It's Darwinism in education, where only the strong survive. Except that "strong" doesn't always mean the teachers or leaders with the best skills. Sometimes the ones "selected" by the environment are simply those who can put up with a lot, or those who have limited options for gainful employment. They are suffering, the schools are suffering, and our children are suffering.

Teachers and leaders who enter public education without the prerequisite skills for teaching and leading pose great challenges to districts, as precious human resources that might be focused on deepening pedagogy and leadership must too often remain focused on developing even the most fundamental skills. Unfortunately, many districts don't have the resources or infrastructure to meet the needs of so many unprepared teachers and leaders through their talent development programs. And because very little of the science of human performance is taught to district and school leaders, performance evaluations intended to motivate and help actually hurt and deepen the problem.

Essentially, good people pursuing teaching and leadership are being thrown into a bad system. As the renowned management expert W. Edwards Deming noted, "A bad system will beat a good person every time" (The W. Edwards Deming Institute, n.d.). Focus must be taken off individual performance and placed on improving the system.

Now, we don't want to play the very blame game we frown on. It isn't our intention to point the finger at universities and other preparation programs. It is our wish that accrediting bodies and those advocating for improvement of teacher- and leadership-preparation programs implement the following:

- Place progressively greater focus on performance measures in order to recognize and reinforce programs that have been effective.

- Complete a comprehensive analysis to discover the key components related to the success of effective preparation programs.

- Support unsuccessful preparation programs in embedding these key components into their curriculum and instruction.

The goal should be to increase the number of successful teacher- and leadership-preparation programs so that all teachers and leaders are adequately prepared to meet the needs of our children, and our children are prepared to meet the demands of a global society.

The good news is that it's starting to happen. A call to action is increasingly being made by various researchers and committees to improve teacher and leadership programs (e.g., The Wallace Foundation and the Wing Institute). Moreover, thought leaders across fields continue to clamor for changes in not only how we teach, but also what we teach and how we motivate students.

Dr. Joe Harless, a noted behavior scientist and advocate for evidence-based educational reform, called for a radical change in our country's approach to education in his book *The Eden Conspiracy* (2017). In it, he advocates for a bottom-up approach that determines the purpose and goals of education and works backward from there. An expert in performance engineering, he outlined this progression with examples for illustration (Figure 3.2).

Figure 3.2. Dr. Joe Harless's bottom-up approach to education reform illustrates the alignment of goals, accomplishments, behavioral processes, and inputs, all driven by the end purpose of creating accomplished citizens. Adapted from Harless (2017).

Harless's hard work may not have gone unheard, as legislation has been popping up that attempts to align the needs of the workforce with the inputs provided by state education and universities. At the state level, forward-thinking legislators have been working toward change. For example, in 2018, lawmakers in Ohio tried to pass House Bill 512 to merge the state education, higher education, and workforce departments into one (Ohio H. R., 2018). Proponents of the bill cited the large number of graduates who require remedial classes upon graduation as a need for the overhaul. Though met with opposition by the state school board and stalled in committee, the purported goal of House Bill 512 was to have better alignment between these systems so that graduates would be equipped with skills that better match those sought by colleges and employers.

And at the federal level, the Department of Education (n.d.-b) originally announced regulations that require states to report on and rate teacher-preparation programs using a number of measures, including:

- student learning outcomes
- placement and retention rates
- feedback from graduates on the program
- feedback from employers on the program

But this still hasn't done the job as the U.S. Departments of Education and Labor (U.S. Department of Education, 2023a) recently announced initiatives to advance teacher preparation programs and expand Registered Apprenticeships for educators. These efforts are part of the Raise the Bar: Lead the World initiative aimed at improving learning conditions and addressing educator shortages. The announcement includes new National Guidelines for Apprenticeship Standards for K–12 teachers, over $27 million in funding to support educator preparation programs, and more than $65 million to develop apprenticeship programs in education across 45 states.

While these regulations may have been met with ire from some higher education institutes, the robust conversations they've generated are important to all stakeholders. As our world continues to change, so do the learning needs of our students. As such, the need for teaching practices to keep evolving is logical. We must continue to design and realign preparation programs with the needs of students and schools if we are to allow them to meet the demands of global competitiveness.

And then we must ensure that required knowledge and skills be effectively transferred to new and existing teachers and leaders. This can be done by regularly reviewing and establishing objectives, analyzing tasks, identifying the most important standards, assessing teachers' and leaders' current repertoire, providing training in critical skills, and then *deliberately* coaching based on their needs.

Where Do We Go?

So why is this important, and what does it have to do with this book? You've come to read it because you are a leader, coach, consultant, or stakeholder invested in supporting teacher success. The problem with some folks who support teachers is that they tend to end up playing the blame game too. And if you understand what's at the root of performance deficits, you might be more understanding and supportive. For decades, focus has been placed on telling teachers what they should be doing and providing poor evaluations when they don't do it.

If we want students to achieve, we must improve our methods of training and coaching teachers. Instead of telling, we must focus on helping. Even now, as teachers enter the workforce without adequate preparation, we must fight to support them. And leaders too! One of the advantages of this book is that the coaching process we outline can be used at any level within the school system. Whether we're talking about district leaders supporting school leaders or school leaders and coaches supporting classroom leaders, *Deliberate Coaching* can enable people to reach the highest levels of success through a measurable process that progressively helps educators perform a little bit better tomorrow than they did today.

4

Use the Wheel, Don't Reinvent It!

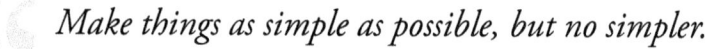 *Make things as simple as possible, but no simpler.*

—ALBERT EINSTEIN

When it's time for a new set of tires, a driver doesn't attempt to design a new tire. Instead, she goes to the local tire store, peruses the assortment of tires, and chooses the best tire for her car in her budget. While all of these available tires are made by different brands, they consist of the same basic combinations of materials that make up a tread and a body. These materials are developed and tested by scientists and regulated by agencies such as the United States Department of Transportation. No person in their right mind would try a "flavor of the month" tire developed by the plumber around the corner as they set off on a cross-country trip. That would be silly and dangerous.

And yet, this flavor-of-the-month approach has existed in education for decades. It is silly. And it is dangerous. It's also unnecessary, as a heavily researched process has already led to positive and sustainable outcomes across many areas (see Critchfield et al., 2023). These areas include, but are not limited to:

- animal training
- autism spectrum disorder
- behavioral medicine and health
- economics
- forensics
- gambling
- gerontology
- marketing
- mental health disorders
- organizational cultural change
- parenting
- physical health
- safety
- schools and classrooms
- sports

How can one approach result in such a positive impact across so many fields? In a word: science—specifically, the science of human behavior (behavior analysis). Like the engineering science that led to the development of tires, which allow us to safely move from point A to point B, the science of human behavior has led to the development of processes that help people move toward meaningful outcomes.

You might be thinking, "What the heck does science have to do with coaching in education?" Well, as you likely now understand, science is *exactly* what education needs if we are going to make predictable, replicable, and sustainable change. Science has literally changed our world and the way we interact with it. It has improved medicine, communication, access to knowledge, and travel, just to name a few examples.

And science is poised to tackle many other issues that ail the human condition. Issues related to crime, poverty, education, and more can all be tackled by science. But not just any science, Applied Behavior Analysis (ABA)—the science of human behavior whose purpose is improving socially significant behaviors and enhancing well-being. During the more than 60 years of its existence, ABA has helped improve lives through careful analysis and intervention. It has also informed policy, programs, and interventions of groups and organizations. And it can help solve complex global issues as well (Ledoux, 2014). This makes sense given that every person in the world can produce better outcomes through technology rooted in ABA. This includes educators and the systems that support them.

Behavior = Results

In a system as large and complex as education, it's easy to get caught up in the myriad of problems and look for the quick flavor-of-the-month fix. The politicians say, "Do this," the economists say, "Do that," and still countless other professionals tell educators, "Do the other." Then the consultant declares, "You have to do one more thing." And what's crazy is that most of them don't seem to care what professionals within the field of education think. It's enough to make your head spin.

Well, here's the good news. In a sense, they're all recommending the same thing: that something be done. And at the most foundational level, they're correct. You see, all results require somebody to do something more, to do something less, or to do something differently. To put it simply, all results require behavior. The common variable among all the recommendations is that they involve people behaving. In education, different behaviors by different people lead to different results.

For example, teachers instruct (this is behavior) in ways that lead to student learning (learning outcomes are a measure of behavior). Principals lead (leading is behavior) in a way that affects a school culture (culture is shared behavior). So it stands to reason that if we want to achieve desired results in education, we must first understand the principles of behavior, then identify the right behaviors that will get results in the right way.

We'll talk more about identifying the right behaviors later on. But first, let's take a minute to talk about right and wrong behaviors. When we say "right," we mean behaviors that are ethical and lead to positive outcomes as defined by a person and the system. There are many examples of results being obtained the wrong way in education. One need only google "cheating on standardized tests" for a long list. From students and teachers to school and even district leaders, these folks sought to achieve the right results via the wrong behavior. This is why it's critical that focus be taken off outcomes such as standardized test scores and placed on the behavior required to achieve intended, valued results.

In the best organizations, these behaviors are strategically aligned to obtain results quickly and efficiently. For example, at the district level, good superintendents collaborate with principals and district leaders to share their vision, identify goals, and identify leadership behaviors that are valued and required to reach those goals. At the school level, good principals collaborate with leadership teams to identify specific goals aligned with the school's vision and district goals, then embed the required behaviors of targeted staff into school improvement plans to see that these goals are accomplished. These principals specify the behaviors they value, then seek to support their leadership team in modeling behaviors that are consistent with these values.

At the classroom level, teachers collectively understand the vision and the behaviors that will lead to its realization. Their approaches to meeting the needs of the students are aligned with the values of the principal and reflect school improvement behaviors and goals outlined in the school improvement plan. Like a well-prepared football team that wins the Super Bowl, this strategic alignment allows schools to accelerate achievement by multiplying key behaviors that lead to established goals.

Oh, and this doesn't exclude you! Your behavior is a critical part of this puzzle. If you know what a teacher should be doing more of, less of, or differently to improve an outcome (such as student achievement), then you must also understand what you have to do (how you should behave) to have a positive impact on the teacher's performance.

Fortunately, behavior analysts have stockpiled mountains of literature aimed at just that. This literature will show you how to arrange the environment (you are a part of a teacher's environment) in a way that brings out the best in people. Understanding why certain behaviors occur under certain circumstances and why other behaviors don't will provide you with important knowledge for influencing those you support to choose more of the right behaviors under the right conditions.

Revisiting No Child Left Behind

In revisiting NCLB and its successor, the Every Student Succeeds Act (ESSA), it's crucial to evaluate these legislative frameworks through the lens of behavioral principles. While NCLB was replaced by ESSA in 2015, the core issues related to the application of behavioral science in educational reform persist. Both policies, despite their differences, seem to lack a foundation in the essential elements of behavioral science, which can significantly impact educational outcomes. We encourage you to review U.S. Department of Education (n.d.-a) resources for more on these two laws.

In Chapter 1, we discussed the concept of metacontingencies, highlighting the importance of interdependent behaviors within an educational system for producing collective outcomes (Malott, 2003). Both NCLB and ESSA, despite their attempts to reform education, do not fully leverage this understanding of collective behavioral dynamics. The legislation under NCLB set ambitious goals, such as achieving 100% student proficiency by 2014, without adequately considering the varying starting points and capacities of individual schools and students. Setting goals without sufficient detail may result in expectations that are too high and objectives that cannot be realistically achieved (Glenn, 1988, 1991). This approach can hinder progress by aiming for targets that are beyond reach, emphasizing the importance of detailed and attainable goal setting in research and practice (Johnston & Pennypacker, 2009).

ESSA, while offering more flexibility and state-level control compared to NCLB, still requires a critical examination from a behavioral standpoint. It mandates states to measure performance in core academic areas and to develop state report cards for transparency (Every Student Succeeds Act, 2015). However, the effectiveness of these measures can be questioned if they do not align with the principles of behavioral science, such as individualized goal setting, precise measurement, and contextual understanding of student and educator behaviors.

The essence of effective educational reform lies not only in setting standards and measuring outcomes but also in understanding and influencing the behaviors that lead to these outcomes. This requires a shift from a purely standards-based approach to one that incorporates a nuanced understanding of human behavior, motivation, and the environmental factors that influence learning and teaching.

Integrating behavioral concepts such as pinpointing, shaping, feedback, and motivation into ESSA could significantly enhance its effectiveness. Let's explore how each concept could be applied to improve ESSA.

1. **Pinpointing:** This involves identifying specific, measurable behaviors linked to desired results. In the context of ESSA, pinpointing can improve the act's focus on individual school performance. Instead of broad, nationwide metrics, pinpointing allows for setting specific behaviors linked to achievable goals for each school, considering their unique contexts and challenges. This approach ensures that performance management is not solely results-focused but also behaviorally informed, leading to more tailored and effective educational strategies.

2. **Shaping:** Shaping focuses on reinforcing incremental steps toward goals, which is crucial in education, where progress often occurs in small

increments. By incorporating shaping into ESSA, schools would be encouraged to recognize and reinforce small successes in teacher performance as they relate to student success. This approach supports ongoing progress rather than waiting for perfect outcomes, thus fostering a more positive and productive learning environment.

3. **Feedback:** Effective feedback is critical for learning and performance improvement. Under ESSA, integrating systematic feedback mechanisms could enhance teaching and student learning experiences. Feedback should be timely, specific, and directly related to the behaviors and goals being targeted. This approach helps in adjusting strategies and interventions promptly, ensuring that they are always aligned with desired educational outcomes.

4. **Motivation:** Understanding motivation in terms of outcomes and consequences can significantly influence behavior. In applying this to ESSA, policies and practices within schools should be designed to motivate both educators and students by clearly linking actions to meaningful outcomes. This involves creating systems where positive behaviors are reinforced and desired outcomes are clearly communicated and understood.

Incorporating these behavioral principles into ESSA would require a shift from a one-size-fits-all approach to a more nuanced, behaviorally informed strategy. This would involve not only setting goals and measuring outcomes but also understanding and influencing the behaviors that lead to these outcomes. Such an approach aligns with the goals of ESSA to provide high-quality education while respecting the unique needs of different schools and communities.

Before determining how to use these concepts to right the ship and set sail, you must first understand why it's sinking. Fortunately, Organizational Behavior Management, a subfield of ABA with a focus on making a positive difference in the workplace, has practical tools to help diagnose the problem. Think about it as a root cause analysis that determines the factors leading to both productive and nonproductive performance. Once the analysis is completed, you can implement multilevel solutions for behavior, process, and system improvement to make a positive and sustainable impact on achievement (Sasson et al., 2006). Without understanding what is working and what is not, you risk wasting valuable educational time on trial-and-error solutions, potentially stopping something that is working or continuing something that is not working.

To improve performance in education, some of the key questions to ask when looking for your leadership solutions include the following:

- What are the desired results at each level of the organization, and what activities must each individual engage in to obtain these results?

- What variables most strongly influence performance at each level within the school system?

- What variables (e.g., policy, resources, training, and management) might be affecting performance in schools?

- What changes can we make to increase performance to the desired results?

Discovering orderly relations between performance and the system in which it occurs provides the opportunity for meaningful improvement in the critical behaviors that ultimately lead to student achievement.

If we could only go back to a time before NCLB and ESSA to gain an audience with lawmakers, we could guide these well-meaning folks toward developing a system that fosters behavior, routine, and habit change to positively impact student achievement. In a sense, we might be thankful for NCLB, as the failed effort has created opportunity for real change. And although ESSA is still falling short, at least it's a move in the right direction. In the end, for change to be as beneficial and effective as possible, it will need to be purposeful, precise, and systematic—rooted in the application of behavior science.

So we strongly encourage you to use the wheel and not reinvent it. A lot of good science has been conducted and applied over a long period of time to help us make good use of our time as educators and leaders. We don't have to rely on trial and error. We don't have to waste half a school year with approaches that don't work.

This book offers a summary of some key areas of behavior analysis that we feel can benefit your leadership and culture and offers a few tools based on this science. We hope that this knowledge will help save you some time so that you can truly maximize the time you spend with your students.

SECTION 2:

The Science of Behavior

5

The Science of Behavior

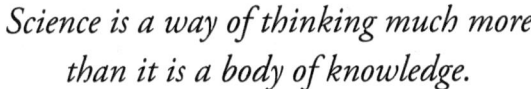

*Science is a way of thinking much more
than it is a body of knowledge.*

—Carl Sagan,
Broca's Brain: Reflections on the Romance of Science

Every school has its own distinctive culture. In the educational arena, the wise school leader sees culture as the hidden playbook, the unsung element shaping every interaction while simultaneously boosting teachers' morale and enabling students to be happy and productive. For this leader, a thriving culture is not a luxury; it's a necessity, serving as the engine that propels both teachers and students to excel. It's their Rocky Balboa, always punching above its weight.

Yet, these leaders are keenly aware that not every school culture can claim this positive status. Some cultures, bogged down by ineffective behaviors and outdated norms, are more of a liability than an asset—like a boxer who's lost his footwork and is taking more hits than he's landing. Perhaps your school is reflective of a strong culture. Then again, perhaps it's not.

But here's the silver lining: Culture isn't set in stone. Why? Because it all boils down to patterns of behavior—behavior of students, staff, teachers, administrators, and others who make your school unique. Thankfully, there is logic behind this behavior grounded in ABA. Why reinvent the wheel when you can learn from what exists and build on it? By doing so, you can avoid some of the costly educational pitfalls we've all faced at one time or another, many of which are highlighted in this book. By harnessing the principles of behavioral science, the school leader or anybody else attempting to improve educational outcomes can adopt a precise, purposeful, and systematic approach for accelerating improvement.

Effective, efficient teacher behaviors lay the groundwork for sustainable student outcomes. These behaviors must be built—one way or another. Exemplar behaviors might occur naturally over long periods of time as a result of trial and error, or they might be trained systematically with a certain level of direction and precision. The purpose of any leadership system is to make it possible to build these behaviors in a way that minimizes the use of school resources and maximizes the impact of school leadership. To do this, we start at the beginning: with our ABCs.

The ABCs of Performance

Every job in an organization has specific behaviors that are critical to its success. A die setter in an injection molding plant must know how to run an injection press, a nurse at a hospital must know how to administer medications, and a teacher must know how to instruct. In order for these behaviors to occur, people receive training, review the organization's procedures, and follow checklists. These antecedents (e.g., training, procedures, checklists, telling) all come before a behavior and serve to get it started or prompt it to occur.

However, whether people will continue to perform on the job is dependent on what happens after the behavior. From a behavior-analytic point of view, this is the elephant in the room. If training, goal setting, planning, and telling were all that were needed to be successful, you probably wouldn't be reading this book. While they are often a critical part of the performance formula, they are insufficient for building and sustaining a culture of success. It is the consequences of behavior that must be focused on to improve performance. In short, if something desired follows a certain behavior, then this behavior will continue. If something undesired follows the behavior, then the behavior will go away. These are the consequences that determine whether a newly trained behavior will continue or will dissipate over time.

All behavior, in and outside of the workplace, follows this antecedent-behavior-consequence sequence—the ABCs (see Figure 5.1). This includes the behavior of our students, teachers, and school leaders.

Let's take a look at a quick example, and then we will dive more deeply into the ABCs.

- **Antecedent**: The teacher receives professional development training on implementing student-centered teaching strategies. She's given resources, lesson plans, and checklists aimed at fostering better student engagement.

- **Behavior**: The teacher starts applying these new strategies in her classroom, creating more interactive lessons and encouraging active student participation.

- **Consequence**: Here's where the real work begins and where permanent change and progress are nurtured. If the school leadership and peers

THE ABCs OF SCHOOL LEADERSHIP

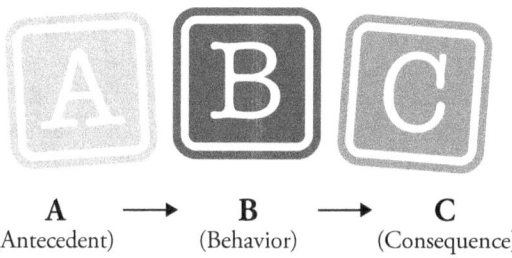

A → B → C
(Antecedent)   (Behavior)   (Consequence)

Figure 5.1. This illustrates the ABC framework (Antecedent-Behavior-Consequence), which outlines how all behavior is shaped by preceding events and followed by reinforcing consequences. This model is critical for understanding and influencing positive changes in settings ranging from classrooms to workplaces.

recognize and praise her efforts—perhaps even providing some sort of other acknowledgment—then that teacher is likely to continue implementing these strategies. This positive reinforcement serves as a potent consequence that sustains the behavior.

But let's say the opposite happens. The teacher tries the new strategies but faces resistance from students used to a more traditional lecture style. Or the school leadership doesn't acknowledge her efforts or even criticizes her for disrupting the usual flow of lessons. These negative consequences could quickly extinguish the new behavior, making the teacher revert to her previous teaching methods.

Do you want your teachers to be good because of you or in spite of you?

In summary, while antecedents such as training and procedures set the stage, they are essentially the opening act. The real star of the show is the consequence that follows the behavior. And this isn't just relevant for teachers; it's the same story for school administrators, nurses, die setters in injection molding plants, even your behaviors in your everyday home life. If you want sustained excellence in your school, understanding and strategically planning for the ABCs is not just a good idea—it's a mandate.

YOU GOTTA FIND IT FIRST

The first step in helping a teacher be successful is to figure out what the desired teaching behaviors are. For instance, what are the student outcomes each teacher is responsible for, and what is the best path to those outcomes? You can't assume that a new teacher comes in with these best practices simply because they have a college degree, nor can you assume that teachers who get good student outcomes necessarily know exactly what they're doing to get those outcomes. It's common to hear things such as "I just understand the students" or "I like to empower the students" or "I just have a good group of students." None of these should give you confidence that the teacher knows what's working and what isn't. This leaves teachers identifying themselves as people with "good personalities," "passion," "charisma," and other vague descriptors of behavior hoping they'll continue to capture lightning in a bottle—and the rest of us are left wishing that we had a different "personality."

But it's not about personality. People aren't born good teachers. Good teaching behaviors such as responding frequently and positively to students are built over time because certain things work, and certain things don't. Good school leaders find out what these behaviors are and then share this information in the form of antecedents (that is, training) to make sure incoming teachers know what to do, and how to do it.

START IT UP

Once you have identified the outcomes of good teaching and the behaviors teachers should engage in to get to those outcomes, the next step is to make sure they can actually demonstrate these best practices. As we stated, antecedents come before behavior and serve to initiate effective behavior. You provide teachers with instructions, so they know the procedure. You provide them with training to build new skills to add to their repertoire, making them more capable of producing successful students in the classroom. You send people reminders about what to do and when to do it.

The science tells us that behavior occurs because something prompts it to occur, whether we're aware of what that something is or not. By understanding this, school leaders can set teachers up for success by building in these antecedents and maximizing the use of the staff and resources to guarantee that the right teaching is occurring in and out of the classroom.

An important consideration when using antecedents, however, is not to rely too much on them. Antecedents can be a fast way to check the box that you've addressed some sort of leadership need in your school. Teachers appear to be struggling with an area, so you fix a policy, provide training, or send out a memo, then move on. But these actions are just part of the equation. If all you do is get a behavior started—the teachers demonstrate best practices in training, and they're starting to apply them in the learning environments—it will be likely to eventually cease if you stop there.

So what's the next step? Consequences!

"THERE WILL BE CONSEQUENCES"

Consequences determine if behavior will continue. Although the word "consequence" might leave a bad taste in your mouth, fear not. Consequences can be good or bad. They can be delivered by someone (e.g., praise, criticism) or occur naturally in the environment (e.g., try something new and save time, giving a ton of effort on a new task). Perhaps the consequence that keeps a teacher's behavior going is the success of a student on a test, or perhaps it's recognition by a parent of out-of-classroom accomplishments. And sometimes it's that the teacher's job got a little easier! It's all dependent on the individual. There are also consequences that stop behavior, which we will cover as well.

Reinforcing Consequences

It's easy to see that if something follows a behavior and the behavior increases, this means the person finds some sort of value in that consequence. We call this positive reinforcement. A teacher who feels rewarded when students get a correct answer will engage in whatever behaviors result in students getting correct answers. A teacher who likes to receive praise from the principal will engage in whatever behaviors have led to praise in the past.

The term "positive" often gets misinterpreted, and it's a crucial point to clarify, especially when we're talking about improving behavior. You see, in everyday language, positive usually means "good" or "beneficial." It's a judgment call. But when we step into the scientific arena of behavior analysis, positive takes on a more nuanced definition. Here, it simply means "added." In the example above, when the teacher engages in the specific behaviors that lead to students getting correct answers and finds value in that outcome, it's not positive because it's morally good (although many would argue it is), but it is positive because something has been added—the value or pleasure derived from seeing students succeed.

It's important to understand that when reinforcement stops, the behavior goes away. For example, if the students stop getting good grades, or the principal stops providing praise, the teacher is no longer having consequences that encourage him to continue the behavior. This process of behavior eventually stopping because it's not being reinforced is called extinction. In other words, much like dinosaurs disappeared or became extinct because there wasn't an environment to sustain them, behavior will eventually disappear too without a supportive or reinforcing environment.

In contrast to positive reinforcement, if something is removed or avoided following a behavior and the behavior increases, we call this negative reinforcement. If a teacher finds that a classroom activity is time-consuming and difficult, she might avoid that activity but engage in another one. If a teacher does not want to be around the principal, she will do whatever it takes to avoid the principal.

Behavior maintained by negative reinforcement can also be extinguished if the person is no longer able to avoid or escape the bad outcome. For example, a teacher who doesn't like cleaning up before he leaves might ask an aide to stay late and clean. The aide cleaning up for him negatively reinforces (i.e., increases the likelihood the teacher will leave early) leaving before all of his duties are done. Once the aide stops cleaning up after him, the teacher will be forced to stay and clean up for himself.

Let's further explore these two contrasting examples from the lens of school leadership, one involving positive reinforcement and the other leveraging negative reinforcement.

SCHOOL LEADER USING POSITIVE REINFORCEMENT

Meet Principal Johnson, a proponent of positive reinforcement. She makes it a point to visit classrooms regularly to observe teachers in action. Whenever she notices a teacher using an effective instructional strategy or managing the classroom well, she offers immediate, specific praise. She even takes it a step further by acknowledging these achievements in staff meetings and newsletters. She sometimes provides small tokens of appreciation such as gift cards or an extra planning period.

Consequence: Teachers under Principal Johnson's leadership feel valued and acknowledged. They're motivated to continue engaging in effective teaching practices, not just because it's the "right thing to do" but because they know their efforts will be recognized and rewarded. In the long run, this positive reinforcement nurtures a thriving school culture.

SCHOOL LEADER USING NEGATIVE REINFORCEMENT

On the other side, we have Principal Smith, who takes a different approach. He closely monitors his teachers and often highlights areas where they could face potential repercussions, such as low student test scores or negative parent feedback. When a teacher shows improvement in these areas, Principal Smith removes or reduces previously mentioned threats. For instance, he may abstain from placing the teacher on a performance improvement plan if test scores rise.

Consequence: Teachers in this scenario may improve specific behaviors to avoid adverse outcomes. The negative reinforcement—the removal of a looming threat—does result in behavior change, but it could be at the cost of creating a more stressful work environment. Teachers might comply out of fear rather than genuine interest in professional growth.

When teachers such as the one in the above scenario operate under the cloud of negative reinforcement, the immediate consequence might be behavior change, but let's unpack the ripple effects, shall we? Firstly, there's the ever-pervasive feeling of fear or anxiety. Imagine constantly walking on eggshells, uncertain when the next threat of negative repercussions might loom large. This emotional state can be mentally exhausting. In behavioral terms, we're talking about a heightened state of "emotional responding," which can interfere with clear thinking and decision-making.

> *If teachers are solely focused on avoiding negative outcomes, they're less likely to take risks or try new instructional strategies.*

Then there's the stifling of creativity and initiative. If teachers are solely focused on avoiding negative outcomes, they're less likely to take risks or try new instructional strategies. The culture becomes one of mere compliance rather than of true engagement or innovation. In other words, teachers are not so much driving the car as they are desperately trying to avoid a crash.

Let's not forget the issue of collaboration. Teachers may become less willing to share difficulties they're facing or seek help from peers or administrators, fearing that revealing any struggles might put them on the radar for further negative attention. This could lead to an erosion of the professional community, where each teacher becomes an island unto themselves. Also, there's an ethical dimension here: Operating from a standpoint of fear can compromise the integrity of educational objectives. For instance, a teacher worried about test scores might resort to "teaching to the test" rather than aiming for comprehensive understanding and critical thinking.

Lastly, consider the message this approach sends to the students. They can undoubtedly pick up on a teacher's stress or apprehension, and this emotional climate could well affect their own learning and well-being. So, while negative reinforcement might yield some quick fixes in terms of altering behavior, its long-term costs could be steep, affecting not just individual teachers but also the collective ethos of the educational environment. Therefore, a wise school leader would do well to weigh these potential downsides carefully when considering their approach to coaching and management.

So what's the solution? You've got it: positive reinforcement—a coaching strategy that's not just effective but also deeply humanizing. Let's delve into the ripple effects when a school leader leans into this approach. First off, positive reinforcement does wonders for boosting morale. A teacher who feels acknowledged and appreciated is likely to be more engaged and satisfied with their work. In the behavior science lexicon, this is often referred to as "strengthening the behavior-environment relationship."

The second point is about professional development. With positive reinforcement, there's a built-in feedback loop. When a teacher knows exactly what they did right, they're more likely to replicate that behavior, but also to refine it and even innovate.

It's like giving someone not just the fish but also the fishing rod, along with a pat on the back.

Thirdly, there is an impact on collaboration and a sense of community. When positive reinforcement is part of the school culture, it often spills over from administration to faculty, from faculty to students, and even back and forth among faculty members themselves. Teachers are more likely to share best practices, celebrate each other's achievements, and offer constructive support. There's a shared ethos of "We're in this together," which is priceless.

Also, positive reinforcement promotes risk-taking and creativity. With the security of knowing that efforts are recognized and valued, teachers become more willing to try new approaches or strategies. They're not just sticking to the script; they're coauthors in an unfolding educational narrative.

And let's not ignore the modeling aspect. Students aren't just passive observers; they absorb the behavioral norms modeled in their environment. When they see teachers being positively reinforced, it sets an example for them. They learn the value of positive behavior, teamwork, and mutual respect. It's a foundational lesson that will serve them well beyond the classroom walls.

Last but not least, the benefits of positive reinforcement extend to school leaders themselves. The act of giving positive reinforcement can also be reinforcing for the giver. Leaders may find greater job satisfaction in fostering a thriving work environment. This can lead to decreased turnover rates and a stronger, more unified organizational identity. In essence, while the immediate goal of positive reinforcement may be improving behavior, its long-term outcomes can be transformative.

Positive reinforcement promotes risk-taking and creativity. With the security of knowing that efforts are recognized and valued, teachers become more willing to try new approaches or strategies.

Positive reinforcement is not just about changing what people do; it's about enriching who they become in the process. This is leadership that not just instructs but elevates.

In sum, in the examples above, both leaders achieve behavioral change, but the paths they take diverge significantly. Principal Johnson's positive reinforcement strategy adds something valuable to encourage good behavior, fostering a more engaging and supportive culture. In contrast, Principal Smith's negative reinforcement removes something undesirable to bring about compliance, which might achieve short-term goals but could potentially strain the school's culture in the long run. There are also ethical implications that can come into play here. We will talk about that shortly.

The distinction between positive and negative reinforcement is important when you think about the type of coach you are and the type of coach you want to be. Do you only show up when something is wrong? If so, consider the effect this pattern of behavior has on your school culture. If something good happens in the classroom, nobody knows about it and the teacher is ignored. If something bad happens, a school administrator shows up. With these as the sole consequences, it's easy to see why some teachers will do just enough to avoid an uncomfortable situation—the uncomfortable situation being an encounter with you!

Do you want your teachers to be good because of you or in spite of you?

Punishing Consequences

Now, we mentioned behavior is a function of its consequences. But we have yet to mention another type of consequence that impacts behavior: punishment. From a behavior-analytic point of view, punishment is often misunderstood. It's not about casting blame or chastising; it's simply a consequence that decreases the frequency of a particular behavior. It should never be about punishing people; it should focus only on correcting errors in behavior. For example, consider a scenario where a school leader observes that a teacher is consistently tardy with grade submissions. Instead of reprimanding the teacher, the leader kindly requests punctuality from the teacher, emphasizing the importance of deadlines for student feedback and administrative processes. Should the teacher subsequently improve their timeliness, the polite request would serve as a punisher for the tardy behavior, effectively decreasing its occurrence.

There was nothing coercive or inappropriate about this interaction. It was simply information politely provided to the teacher that stopped their tardy behavior. After this change in behavior, the school leader should then focus on using positive reinforcement techniques like thanking the teacher or sending appreciative private emails for on-time behavior. In some cases, leaders can even give public shout-outs to all faculty and staff as a group for increases in on-time behavior. For example, "I want to give a huge shout-out to our faculty and staff. I noticed almost all of you were on time at your posts this morning, ready to welcome our students. And the ones who weren't on time were able to give me a heads-up so I could be sure to fill their spot until they could make it." If improving on-time behavior is a focus, then initially, this reinforcement is best delivered consistently to establish the behavior firmly—for example, every deadline met results in positive feedback. Every morning that faculty and staff are at their posts, they receive a positive email thanking them.

However, leaders equipped with a toolbox made up of the science of human behavior understand the power of reinforcement schedules. Once the desired behavior becomes the norm, the leader can begin to fade the continuous reinforcement. Instead of an appreciative email for every deadline met, it might come every third or fourth time. This is intermittent reinforcement, a schedule that's known to be very effective in maintaining established behaviors over the long term.

The beautiful thing about intermittent reinforcement is that it doesn't just stop with one teacher; it's a strategy that can and should be generalized across staff. By varying the times and instances where positive reinforcement occurs, the leader fosters an environment where good behavior doesn't just appear; it endures. This nuanced approach ensures that the faculty and staff don't become dependent on constant praise to perform well but also feel valued and acknowledged sufficiently to maintain high levels of performance.

When it comes to positively reinforcing versus correcting (punishing) behavior, the golden rule to remember is a 4:1 ratio—four positive interactions for every corrective one (Madsen & Madsen, 1974). It's not merely about stopping behaviors we don't want to see; it's about abundantly fueling the behaviors that will help to achieve valued outcomes. This balanced approach promotes a healthy, thriving environment where people feel valued and are motivated to grow. In an educational setting where turnover is plaguing the system, such an atmosphere creates a positive learning culture characterized by high retention and happy people. In contrast, even in the most well-intended hands, the misuse and overuse of punishment can have a number of unfortunate outcomes. For example:

- An environment where punishment is heavily used can become tense, fear-ridden, and stressful. Students and teachers might feel anxious about making mistakes or taking risks, which hinders creativity and engagement. This negative emotional climate can also impact overall well-being and morale.

- Punishment alone doesn't teach individuals what they should be doing instead. It only suppresses unwanted behavior temporarily. Without clear guidance and reinforcement for positive behaviors, students and teachers may not know how to meet expectations or perform better.

- Individuals subjected to frequent punishment may begin to engage in behaviors aimed at avoiding punishment rather than actively seeking positive outcomes. This can lead to a focus on staying out of trouble rather than a drive to actively pursue excellence.

- Overuse of punishment can foster resentment toward authority figures and create resistance to change. People might comply with expectations out of fear rather than a genuine desire to improve, leading to superficial adherence rather than meaningful growth.

- A punitive environment discourages risk-taking and innovation. Students and teachers might hesitate to try new approaches or strategies, fearing negative consequences for any missteps.

- Prolonged exposure to punishment without adequate positive reinforcement can inadvertently reinforce negative behavior patterns. Individuals might believe that their efforts go unnoticed or unappreciated, leading them to revert to familiar but ineffective behaviors.

- Frequent punishment can strain relationships between educators, students, and administrators. Trust erodes when punishment is perceived as unfair or inconsistent, leading to a breakdown in communication and collaboration.

- Continuous punishment diminishes motivation. Students and teachers might feel coerced into compliance rather than driven by the natural consequences of their behavior to achieve their best. For example, reading for the sake of reading as opposed to reading for fear of a consequence.

- An emphasis on punishment doesn't provide the necessary skill development for students or teachers. Effective learning and growth require clear guidance, constructive feedback, and opportunities for skill-building.

- An environment dominated by punishment can contribute to high turnover rates among both students and teachers. People are less likely to remain in an environment where they feel constantly criticized or undervalued.

- Punishment often provides immediate results, but it's not effective for long-term behavior change. Over time, individuals might become desensitized to punishment, rendering it less effective in influencing behavior.

In essence, finding the sweet spot between the application of positive reinforcement and punishment is paramount. Positive reinforcement takes on the role of the motivating engine, nurturing a sense of purpose, belonging, and inner drive. At the same time, the strategic application of corrective measures, or punishment, serves as a guiding tool that nudges behavior and performance toward enhancement. Through the seamless integration of these strategies at rates of a minimum of 4:1, we create a space in which individuals thrive.

Getting to the Root of Performance Issues

Most teachers chose their career path because they genuinely care about children. They want to see their students succeed. The right teaching behaviors should lead to the right student outcomes. However, student outcomes aren't the only consequence of teaching. Teaching is hard. It takes effort. Doing it the right way can be time-consuming, and it's often a thankless job, with teachers only hearing from parents and administrators when something goes wrong. As a school leader, you can help by adding positive consequences—making good things happen more often for the right behaviors. There's a lot you can do by yourself in the form of praise, feedback, and other tenets of a good coaching system. And if a certain teacher has performance problems, you need to be sensitive to all the antecedents and consequences that surround this teacher's day-to-day performance.

In the case of a performance deficit, you can build better behaviors through the process of shaping. Shaping is much like creating a sculpture. When molding clay, the sculptor has a clear idea of what she wants it to look like, but she certainly doesn't start off with a perfect sculpture. She starts with a lump, formless and malleable. With each touch, squeeze, and adjustment, the clay begins to take the shape she intends. If she presses too hard or too quickly, she risks making an irreparable dent. If she's not deliberate, the shape remains vague and unformed.

Similarly, in the world of ABA, shaping behavior is an incremental process. You start with existing behaviors, which might be as unformed as that lump of clay, and through a systematic application of antecedents and consequences, you guide these behaviors toward a desired outcome. Whether it's a teacher integrating a new classroom management technique or an employee learning a complex skill, each small, successful step is acknowledged and reinforced, nudging the teacher or employee closer to the final form of desired behavior.

Just as you wouldn't expect to create a perfect clay sculpture in one fell swoop, you can't expect flawless behavior instantaneously. Both processes require patience, precision, and a keen eye for recognizing when the shape—be it clay or behavior—is veering off course. Corrections are gentle but deliberate, made with the vision of the end goal firmly in mind.

In both cases, it's an ongoing process of adding a bit here, smoothing a bit there, reinforcing progress, and making mindful adjustments. And the true artistry lies in knowing how to apply just the right amount of pressure or reinforcement at just the right time to bring the envisioned form to life.

To shape, you simply identify the overall goal you want the teacher to achieve, figure out what the teacher is doing now, and reinforce incremental improvements toward that goal. For example, if you want a new preschool teacher to have the type of jovial, energetic interactions most seasoned teachers have with their students, you have to start where the teacher currently is. If the new teacher barely smiles and only communicates with a few of the children, you can't expect her to jump to your standards overnight. And if you simply choose to wait for her to do it, maybe providing more training because you assume she doesn't know how, you'll be waiting a long time.

So you start small. Reinforce her simply talking more to the kids. Once she's doing that, start focusing your reinforcement on how she talks to the kids. Is she smiling? Is she laughing? Is she engaging the students with questions and pretend play? Slowly move toward the goal and reinforce her improvements along the way. This is the path to success for a good coach, and this is the way to maximize staff performance while building a positive school culture.

The teaching behaviors that have the strongest antecedents and the strongest consequences win. This means that if there are powerful antecedents pushing poor teaching and powerful consequences keeping those behaviors going, best practices will lose out to these less desirable methods. It's not necessarily because the teacher can't teach the right way or doesn't want to teach the right way. The school leader's job is to find out what's causing optimal and suboptimal teaching, make sure teachers have the skills, and use antecedents and consequences to refine and maintain the desired behaviors—show them that teaching the right way pays off for everyone.

The Ethics of Coaching

School leaders and all those charged with supporting teacher performance have likely been extensively trained in their craft. You probably had a very successful career as a teacher, with a proven track record of getting results, prior to taking responsibility for the performance of instructors who work to maintain the high standards you once set. The scope of your day-to-day work has drastically expanded—you've moved from classroom- and student-based activities to administration and staff management. Now you must lead your teachers in a way that generates the teacher performance that yields the best outcomes for each student, each classroom, and the school as a whole.

Although ethics is a broad concept, the central premise is this: getting results the right way. What is the right way in this case?

- by consistently supporting teachers rather than waiting for something to go wrong and popping in to let them know about it

- by helping teachers rather than simply evaluating them

- by building behaviors through positive reinforcement rather than relying solely on punitive measures for performance problems

- and, perhaps most importantly, by focusing not just on school and classroom results, but on how those results are attained

An initial requirement of new leaders, regardless of industry, is to identify their new responsibilities and assess their own strengths and weaknesses. However, even if you have no prior training or supervised experience leading others, the basic principles (ABCs) of behavior that were responsible for your success as a teacher still apply; they just need to be redirected toward building a workforce of high-performing teachers instead of a classroom of high-performing students. Positive reinforcement can accelerate teacher performance just as it did for your students when you used it in the classroom. You just need to be able to take a bird's-eye view of your role and see how you can expand your understanding of learning and performance to a group with vastly different needs.

In behavior analysis at the time of writing, this ethical compass is embodied in the *Ethics Code for Behavior Analysts* established by the Behavior Analyst Certification Board® (BACB®) (2020). This code illuminates the path for behavior analysts, ensuring their practices enrich lives without crossing the boundaries of dignity and respect. Every behavior analyst under the BACB's umbrella pledges to adhere to these guidelines, embedding ethical considerations into the very fabric of their work.

The use of behavior analysis in schools, as outlined in this book, emphasizes a scientific approach to understanding and influencing human behavior, offering school leaders (and others) a powerful framework for coaching staff and improving student engagement and achievement. Through behavior analysis, school leaders can develop evidence-based strategies that reinforce positive behaviors, address challenges proactively, and lead with an informed, data-driven mindset.

The ethical dimensions of behavior analysis further enrich this approach, providing a moral compass that guides interventions and coaching practices. Just as school leaders are committed to the welfare and development of their students and staff, behavior analysts operate under a code of ethics that prioritizes respect, dignity, and the promotion of beneficial outcomes. This shared ethical foundation underscores the alignment between behavior analysis and effective school leadership, emphasizing responsibility, integrity, and the pursuit of excellence.

Integrating behavior analysis into the coaching equips you with the skills and insights needed to cultivate high-performing teams; enhance instructional practices; and create positive, engaging learning environments. It's about leveraging scientific understanding to make informed decisions and implement strategies that not only resolve current issues but also anticipate and mitigate future challenges. This book aims to bridge the gap between theory and practice, offering you a pathway to apply behavior analysis in your coaching efforts to achieve transformative educational outcomes.

SET THE TEACHER UP FOR SUCCESS

Coaches should consider any environmental conditions and other barriers that might interfere with teacher performance. This helps set teachers up for success and frees them to engage in their most optimal performance. There are a number of items that could hinder an otherwise high-performing teacher. Perhaps the teacher is capable of exemplary performance but simply hasn't been trained (can't demonstrate the skills you're looking for when asked). The teacher's lack of skills is a barrier that you need to address.

Environmental conditions such as a lack of materials, large class sizes, and procedural or scheduling problems can interfere with performance as well. Perhaps a group activity is designed to take 30 minutes to complete, but a teacher is consistently left with only 15 minutes. This suggests that you need to help the teacher figure out why her schedule is consistently off and how to fix it. Perhaps another activity is taking too much time. Maybe time is being eaten up by student disruptions that need to be addressed.

Whatever the issue, it's the job of a good leader to identify the barriers to success.

IT'S NOT JUST ABOUT POOR PERFORMANCE

Coaches are ultimately responsible for generating more good performance, not just less poor performance. If you just focus on stopping poor performance, you'll have a workforce searching for what to do instead. Consider the following example.

> Mountain High has a teacher-management program that consists of large lectures in the auditorium during professional development days, teacher evaluations conducted twice during the year, write-ups for teacher performance problems, and a tenure system. The professional development days involve the teachers gathering to listen to a speaker a few times a year. There are no demonstrations of learning or structured opportunities to engage during the training; it's simply a presentation. While this means that the impact is unknown, training like this isn't likely to build more than a general awareness of the content.

> The teacher evaluations are designed to check in on the teachers, and the measures include subjective items such as "Instructor is creative in their approach" and "Instructor is efficient." But this feedback is delivered infrequently; it usually doesn't come until long after the behaviors have occurred, and it's too subjective to have an effect. Write-ups for teacher infractions including tardiness or other rule violations rely on punishment to decrease behavior; while this isn't necessarily a bad thing, they don't positively reinforce desired behaviors (e.g., showing up on time, following school rules). In addition, overreliance on punishment is likely to have a detrimental impact on culture as it sometimes elicits strong emotional responses from those being punished that can have a negative ripple effect across the school. Finally, while the tenure system provides job security, it doesn't recognize teacher performance.

Overall, Mountain High's system involves infrequent contact with teachers, thus it provides limited support for them. It relies on ineffective evaluation procedures and more punitive measures, which can often lead to an adverse climate and increased rates of teacher attrition. Least-restrictive alternatives start with procedures to support the teachers before resorting to more punitive measures. You can't achieve your goals as a leader simply by focusing on decreasing undesired behavior. Results come by people doing the right things, and these right behaviors occur through your use of antecedents (e.g., training, prompts) and reinforcement.

FEEDBACK

We talk a lot about feedback in this book—and for good reason: It's essential and it works! In fact, it might be considered the Swiss Army Knife™ of performance improvement. For example, feedback can:

Strengthen Behavior: Think of feedback as the iron that sharpens iron. When an individual receives specific, positive feedback for a behavior, it's like giving that behavior a vitamin shot. It gets stronger, is more consistent, and occurs more frequently in the future.

Refine Behavior: Feedback serves as a fine-tuning mechanism. Maybe a teacher is on the right track with a new instructional strategy but needs to tweak certain aspects. Feedback can guide this precision fine-tuning.

Generalize Behavior: Once a specific behavior is mastered in one context, feedback can help extend this mastery to other scenarios or tasks. It can help an employee or teacher apply an effective approach in various settings or with different populations.

Discriminate Behavior: In the realm of behavior science, discrimination involves learning when to perform a specific behavior and when not to. Feedback can facilitate this discrimination. For instance, if a teacher knows that a particular instructional strategy works well for one subject but not another, feedback can reinforce this understanding.

Maintain Behavior: Even well-established behaviors can fade over time due to satiation or competing contingencies. Timely feedback acts like a maintenance dose, helping to sustain the behavior over long periods.

Chain Complex Behaviors: Feedback can guide the learning of a series of related behaviors. For example, constructive feedback at each step of a multistage project can help an individual learn and link together the necessary skills to complete complex tasks.

Foster Self-Evaluation and Autonomy: Once a behavior becomes internalized, the individual may start self-administering feedback, which fosters autonomy. At this point, the person becomes their own feedback loop, so to speak.

Promote Accountability: Feedback fosters a sense of ownership and accountability. It keeps individuals aligned with organizational objectives, continually drawing a line of sight from their actions to broader outcomes.

Build Resilience: Positive feedback after setbacks can build resilience by reinforcing the notions that mistakes are opportunities for growth and effort is as valued as the outcome.

Cultivate a Positive Culture: Consistent, fair, and constructive feedback can have a systemic effect, helping to cultivate a positive organizational culture where everyone is attuned to continuous improvement.

Feedback, when artfully employed, becomes a dynamic tool that doesn't merely trigger or halt actions but molds, shapes, and perfects them. It's akin to the masterstroke of a painter or the final flourish of a composer—subtle, perhaps, but oh so impactful. Coaches need to provide timely feedback to supervisees and use reinforcement systems that improve their performance. The Deliberate Coaching System is designed to offer an avenue for this feedback. You can't provide feedback unless you know the specific behavior you're targeting from the teacher and why that behavior is important. The feedback conversation needs to include this information so the teacher understands. If not, this isn't likely to have an impact on performance—it's simply a conversation, not a feedback conversation.

Feedback should be paired with something of value to the teacher, such as praise or showing them how the performance described in the feedback fixes a personal concern they're having in their day-to-day work (natural reinforcers such as saving time or seeing the progress in students). These natural reinforcers can be very powerful as they do not require someone else to deliver them, thus they often occur more immediately and more frequently than those reinforcers that coaches deliver. This helps feedback to function as a reinforcer and actually improve performance.

Constructive feedback that identifies poor performance and includes a discussion of the replacement behavior needs to be followed up with an opportunity to positively reinforce that replacement behavior. If it isn't, the teacher might go through all the effort of doing the new behavior without seeing the benefit of that behavior. Feedback isn't necessarily reinforcement, but it can be. Consistent feedback can have a significant impact on performance, and it's critical for all leaders.

Again, these principles are generalizable to any area of leadership. The principles of behavior analysis are in effect whether you're addressing a student's behavior, a teacher's behavior, or a coach's behavior. If behavior is occurring, it's occurring because of positive reinforcement or negative reinforcement (e.g., avoidance). If behavior is not occurring, it's because there are more contingencies of punishment keeping the behavior down than reinforcers keeping it going. As a result, behavior analysis is applicable to any person responsible for behavior change—including school leaders.

Ethical coaching means focusing not just on the outcomes, but how you get those outcomes. You don't want teachers doing the bare minimum to get by, and you don't want a culture where people do whatever it takes to get results, no matter how questionable those practices might be.

If you're a teacher, psychologist, counselor, or other kind of school professional, you might abide by your own ethics code, and we strongly encourage you to continue to do so. These professional and ethical standards are put in place to protect you and those with whom you work. Our hope is that their similarity to the ethics of behavior analysis will lead to a continuously developing ethical repertoire that you and your colleagues will value.

Building Habits

We've talked a bit about natural reinforcement, but its importance to the long-term success of a coaching system cannot be overstated. Let's take a look at the concept through a musical example.

Learning to play an instrument can be a time-consuming, difficult, yet rewarding process. You might fool around on the piano and attempt to teach yourself, ready to see whether your inevitable talent will result in your being a jazz prodigy or more of a rock star! After a few failed attempts at "Chopsticks," you'll either give it up or seek some lessons. But the lessons are no easy feat. There is a lot of study, a lot of time with instructors, and countless hours of missed keys and being off tempo that drive you and your neighbors mad. Yet you continue. The instructor is taking it slow, providing encouragement for each new note combination and new section of a piece. Your family and friends are supportive, smiling and supporting you as though you're playing a beautiful ballad. And you need that support at a high level early on.

Now fast-forward a bit. You've learned a piece or two. The compliments start to sound more genuine, and people actually seem to be enjoying your music! But you're also seeking those reinforcers less and playing more on your own. You're liking what you hear when you play, so you keep playing. You're building habits. You're playing music not because you have to, not because others are encouraging you—you're doing it because it's fun and you want to!

Habits are established when you access the natural reinforcers in your environment. Natural consequences occur all the time, and they're all around you. Some of them are good, like the reinforcing sound of music. Some are bad, like the amount of effort it took to get that music to sound good!

One of the reasons natural consequences are so valuable, particularly natural reinforcers, is that they don't require anyone to deliver them. They are consistent and often occur immediately after the behavior. You hit the key on the piano and you hear the note. When that note is bad, it punishes your hitting that key. When that note is good, it reinforces your hitting that key. But if the only thing controlling your behavior is natural consequences when you're just starting out, the frequent, powerfully annoying sound of hitting the wrong keys will vastly overpower the infrequent sound of hitting the right keys.

So why do you keep practicing? You keep practicing because you get a significant amount of external reinforcement that maintains this behavior while you're starting to access the natural reinforcers. This is the job of a coach—to make the connection between improvements in behavior and natural consequences.

Let's take this back to school. When you introduce a new concept to any workplace, it's typically going to involve a bit of a learning curve. Learning will take a lot of time and effort until the person doing the learning gets comfortable with what is being taught.

Say you want a teacher to increase student engagement during activities by asking more questions and providing more detailed feedback to students to prompt continued responses. Prior to your intervention, the teacher told you that lessons have been getting harder to run, students are acting out, and the test scores of the students in his classroom are low. But he has his own history of reinforcement, things that he's done for years because he feels they're working. If he didn't feel there was a reason to run his activities in a particular way, then he wouldn't be doing it.

People do what they do because it works for them, or because it has worked for them in the past. Perhaps their behaviors aren't producing the optimal outcomes, but to them, it's working! This is what you're competing with when you introduce a new requirement.

Now, the coach is about to introduce a new method. The teacher understands and is ready to go. He tries to change his ways to match expectations, but he's struggling and students are noticing. Lessons seem to be getting harder and more confusing, and students are doing worse on the tests and are starting to get upset. So the coach provides feedback and increased support, pointing out that he's getting a lot of students responding, but he needs to tweak how he's providing feedback. After a while, he gets the hang of things and it's paying off. Students are performing better, and his days are getting easier. After a while, the coaching support moves toward other things the teacher is doing, checking back from time to time to reinforce the continued progress he's making with running those activities.

What happened in that classroom? At first, the natural consequences of high effort and poor results would have kept the teacher from doing what the coach asked, if the coach and the principal had not jumped in for continued support. After a while, the new method starts to work—and the coach makes sure the teacher knows it. This demonstrates the value in the coaching relationship and shows that good things happen for the teacher when he works with his coach on improvements.

The goal of coaching is to build good workplace habits. Coaches do this by making the link between coaching suggestions and natural reinforcement. This type of effective coaching is how you build exemplary teachers who can see the impact of their improvements. And this is how you, as a coach, can use the limited time you have to work on multiple behaviors during a school year. You can't do frequent and consistent coaching on the same behavior multiple times per week throughout the entire year. At some point, you'll run out of time and you'll need to support the teacher on other things. But if you simply cut off the reinforcement you're providing without making that link to natural reinforcement, you're putting the behavior on extinction—decreasing performance because you've cut off the main source of reinforcement.

Until other natural reinforcers can take over, such as seeing that progress saves time and actually works with their students, you need to keep providing external reinforcement such as praise and positive feedback to the teacher. Because this can't last forever, it's imperative that coaching conversations constantly help the coachee see the value in their day-to-day work.

Goal Setting

At some point in your coaching planning and conversations, you'll need to collaborate with the teacher to set some goals. Or, if you're a leader coaching one of the coaches, then you'll need to collaborate with them to set goals. How you set these goals can have a big impact on performance.

Consider the notion of a stretch goal, where you set the goal far out of reach. The idea is that people will work extra hard to achieve it, even though it might take superhuman effort. But think about what you've learned about reinforcement and extinction. People do what they do because a positive reinforcer occurs as a result of that behavior. If someone is used to getting a reinforcer and that reinforcer stops, the behavior will decrease (move toward extinction). Consider your boss praising you for staying late for work. It takes a lot of time, but you appreciate that the boss notices and seems to care. If the boss stops this praise, then you might stop staying late because the reinforcement has stopped. If someone never accesses reinforcement in the first place, extinction won't occur (because there's no previous reinforcement to stop); the behavior simply won't get started (e.g., you wouldn't start staying late in the first place). So if the main reinforcer for a behavior is delivered upon reaching an unreachable goal, what type of impact do you think the stretch goal will have on behavior? Not much!

Goals need to be attainable. In fact, coaches should break up their goals. It could take a long time for a teacher to reach an ultimate performance goal, and coaches can't just wait around to reinforce until the end. There is an experimental design in behavior analysis called a changing criterion design. The idea is that you first set a goal slightly above current performance, then raise it a bit once the teacher is hitting the first goal consistently, then raise it a bit more once they are hitting the second goal consistently, and so on. If reinforcement is delivered upon hitting those incremental goals, you have a high rate of reinforcement along the way to achieving the final goal. You do this instead of simply waiting until the individual hits an attainable yet faraway goal.

For example, say you're training for a marathon. The coach doesn't start by setting a goal for you to run 26 miles. The coach starts by saying that the goal for your long run in the first couple of weeks is 5 miles. After you hit that goal consistently, you can move to 7. Once you're comfortable with 7, the goal moves to 10. You don't hit mile 26 until the end of your training, but your running is still being reinforced along the way for hitting these subgoals.

Now consider goals you might set for your teachers. Perhaps a teacher, in a given 15-minute activity, only gets responses from 5% of her students. You've tracked this for some time, and it seems to be consistent. This is the teacher's baseline level of performance, where she's at without your intervening. Most teachers can get at least 80% of their students interacting during a 15-minute group activity. So, do you start by setting the goal for that teacher at 80%? Probably not. Unless you've identified a pretty simple fix for why her rate is so low, most likely this change will take a bit of time. The teacher might need to change a few things she's doing. And remember, she's doing what she's currently doing for a reason. Something about it is working for her based on her current skill set, even if it's not working for the students!

If it's not student engagement, it might be that the procedure is easy or the teacher just simply doesn't know what else to do. You have to find out what's going on and plan to build the alternative behaviors until you finally reach the goal. You know the rate at which you can increase the goal by tracking performance. If you want the teacher to provide all students with a chance to respond during a given activity, but she is currently only getting to 20% of them, then start with a goal of 30%.

Start small with the goal. Change can be hard, and you want to support the teacher by working with her to find ways to succeed and providing a lot of positive reinforcement, especially early on. Once you've identified the reason for the low performance and started to provide the support the teacher needs, maybe you start at 7% or 10%. Show the teacher that you notice the changes in performance, make sure she recognizes these changes, and ensure she knows these changes will be reinforced. There's nothing worse than spending a long time doing things a certain way, being asked to change, and feeling ignored once you make the change. Setting subgoals helps provide a reinforcing path to success.

How you use reinforcement in conjunction with the goals you set will affect how well the goal works to improve performance. Sometimes, simply hitting the goal can function as a reinforcer. But remember that we say "can function" because you only know if something is actually a reinforcer when behavior increases after it occurs. One of the many benefits of tracking performance is that it gives coaches something objective to discuss with the teacher. "You're currently getting through about three lessons." That's a clear, reliable measure that you can get agreement on after your observation of a teacher. Next week when you observe, you can tell the teacher that he got through five lessons in the time he previously typically only got through three. Sometimes just seeing the progress can help, especially when the teacher sees that you're acknowledging it.

There's nothing worse than spending a long time doing things a certain way, being asked to change, and feeling ignored once you make the change. Setting subgoals helps provide a reinforcing path to success.

But there are ways that you can raise how reinforcing it might be for someone to hit a particular goal. You want your coachee to care that they hit the goal, so there are things you can do to help with this. The process of pairing in behavior analysis means that you can take something that has limited-to-no reinforcing value and create a new reinforcer by consistently delivering both the proven reinforcer with the new reinforcer-to-be. For example, going on a walk might be extremely reinforcing to your dog. Before you go for a walk, you might reach for the leash. Eventually, after repeated pairings of the leash and the walk, simply grabbing the leash will cause your dog to go crazy. People work in the same way. Perhaps you've seen that praise from you, a respected supervisor, functions as a reinforcer. If you praise someone for a specific behavior and that behavior increases after you deliver the praise, then praise from you is a reinforcer for this individual. (Note that this doesn't mean praise from everyone will function as a reinforcer.) In this case, you can deliver praise when the teacher hits a goal, and this will eventually make hitting the goal function as a reinforcer even without the accompanying praise.

This pairing process is valuable when you're trying to identify and build reinforcers. You can pair goal setting with any type of reinforcer to raise the value of hitting that goal. If you're tracking performance on a graph or chart, you can also place a goal line on it as a form of graphic feedback. Then the person will see a representation of their performance in relation to the goal, which is another potential opportunity for reinforcement. In the same manner, you can pair yourself with reinforcement so then people are excited for you and your coaching!

If all of the above makes sense, let's look at wrapping up all of these principles and approaches into a nice package that you can work within when goal setting that will drastically increase the likelihood that you can help others achieve goals. Traditionally, SMART (specific, measurable, achievable, relevant, and time-bound) goals (Doran, 1981; Geller, 2003) have been the go-to in education. And while they have their strengths, it is our contention that they leave out critical variables necessary to increase the likelihood that goals will be accomplished. We like to use IMPACT Goals (Gavoni & Costa, 2023).

IMPACT Goals

Goals should give leaders and school staff a clear destination and let them adjust the course as needed. IMPACT Goals are:

- individualized (clear, outcome-focused, and tailored to specific needs and roles of stakeholders, such as teachers dealing with classroom behavioral issues)

- manageable (require providing necessary knowledge, skills, and resources, and focusing on a manageable number of goals at a time)

- positively motivating (involve stakeholder engagement and leaders' actions for motivation, such as involving stakeholders in goal setting and celebrating progress)

- aligned (ensure behaviors required for goal achievement are identified and integrated into daily responsibilities, such as faculty engaging in classroom management strategies)

- connected (focus on short-term accomplishments that indicate progress toward the larger goal, such as developing a classroom management plan)

- trackable (clearly defined and include a monitoring process to measure and evaluate progress, with behavior metrics sorted into leading and lagging indicators)

Let's take a closer look at each element.

INDIVIDUALIZED

The goal should be articulated clearly, defining the exact desired outcome. Moreover, it needs to be tailored to the particular requirements of the individual or group responsible for realizing it. Whether on an individual level or as a group, it's crucial for people to have goals that are pertinent to their specific roles and responsibilities. Take, for instance, a teacher grappling with widespread behavioral issues in the classroom that hinder both teaching and student learning. In this case, a goal centered on enhancing classroom management would be directly relevant and beneficial to addressing the teacher's specific challenges.

- Does the goal state precisely what the desired result is?
- Is the goal specific to the needs of the stakeholders?
- Is the goal relevant to the roles and responsibilities of the stakeholders?

MANAGEABLE

To accomplish goals, people need the knowledge, skills, processes, and resources to be successful. In addition, they can't have too many goals thrown at them at once; otherwise, achieving them is unmanageable. As the old saying goes, nothing is important if everything is important. Therefore, providing stakeholders with knowledge, skills, and resources aimed at just a couple of goals at a time is far more likely to have an impact than attempting to focus on a bunch of goals at once.

- Do staff have the knowledge and skills to reach this goal?

- Do staff have clear processes that guide their behavior (for example, a written classroom management plan with well-established expectations, routines, and procedures)?

- Are resources provided to support those who need to implement or engage in the change (time, tools, money, authority, etc.)?

- Are there a few goals focused on at a time?

POSITIVELY MOTIVATING

The highest IMPACT Goal is motivating stakeholders. We are not only talking about intrinsic motivation, where people are energized simply by the nature of the goal. We are also talking about leaders' actions to motivate people to engage in the change.

Leaders should solicit feedback and involve people in goal setting and planning as a powerful source of motivation. They should then deliberately look for opportunities to reinforce incremental improvement to keep performance moving toward goals positively. The goal needs to be something that stakeholders find value in achieving. For example, in the classroom management scenario, if behavioral challenges occur because teachers report they aren't trained in managing problem behavior, then developing a classroom management plan would likely be a positive, motivating goal.

- Why should stakeholders engage in the change?

- How will feedback be solicited?

- How will behavior be positively reinforced?

- How will progress toward goals be celebrated?

ALIGNED

Goals do not achieve themselves—they require appropriate behavior. Therefore, IMPACT Goals link to the behaviors faculty and staff must engage in to achieve the goal as part of their day-to-day work. For example, an aligned IMPACT Goal for improving classwide behavior in a classroom where behavioral challenges are prominent would be to ensure faculty and staff engage in classroom management strategies taught during training. These might be teaching expectations, observing students, positively reinforcing expected behavior, and quietly and consistently correcting behaviors not aligned with the classroom management expectations. Engaging in the behaviors consistently impacts everybody in the classroom, as more teaching and learning occur when behavioral challenges are minimal.

- Have the specific behaviors required to achieve the goals been identified?

- Is the change outlined and described specifically in the roles and responsibilities of those who will need to change?

- Is this change adding value for the students, teachers, and staff?

CONNECTED

To ensure goals are effectively connected, the focus should be on identifying and valuing short-term accomplishments, not just long-term outcomes. These accomplishments, which will be explored in greater depth in a future chapter, are essentially the immediate results of specific behaviors that act as key indicators of progress toward IMPACT Goals. In the case of improving classroom management, such accomplishments might include the development of a comprehensive classroom management plan, the number of classroom expectations that are prominently displayed, and the extent to which students can correctly recite these expectations. These accomplishments directly contribute to the broader goal of enhancing overall classroom behavior through structured management practices. At both the individual and collective levels within a school, these accomplishments delineate what needs to be done for success, effectively bridging daily actions with the long-term, valued outcomes.

- What accomplishments need to occur for the individual, group, or school to be successful?

- Are the accomplishments aligned with behavior?

- Which of these accomplishments will serve as salient measures of progress toward goals?

TRACKABLE

It is commonly understood that what gets measured moves. For a goal to be measured, it should be described in clear, unambiguous terms and include details of how it will be measured, tracked, and evaluated. A specific goal with a progress monitoring process clearly outlined has a much greater chance of being accomplished because it precisely describes what you are looking to change, at what intervals you will be measuring that change, and what data will measure it. The most effective IMPACT Goals include behavior metrics and are sorted into leading and lagging indicators in the form of accomplishments and subgoals, which we'll unpack in a later chapter. For example, a general goal identified by leadership at a school might be to "improve schoolwide discipline," but a trackable IMPACT Goal is specific and outlined starting with the end in mind.

- What data will be collected?
- How will the data be collected?
- Who will be collecting the data?
- How often will they collect the data?
- How will the data be presented (e.g., graphic form)?
- What is the current level (baseline) of performance?
- Are short-term goals (targets) established?

As we progress through the subsequent chapters, we'll delve deeper into the principles of IMPACT Goals, emphasizing their vital role in educational success. Whether it's to aid a single teacher, a group of educators at a specific grade level, or an entire school, crafting IMPACT Goals is an indispensable antecedent strategy for achieving desired outcomes. The importance of these goals in guiding and shaping the journey toward success cannot be overstated. Detailed and well-thought-out planning significantly enhances the likelihood of reaching these goals.

For coaches, this approach to goal setting is particularly powerful. It allows for the integration of precise pinpointing, rigorous tracking, and effective reinforcement strategies into a comprehensive plan. This plan is not just about setting benchmarks; it's about providing a clear roadmap from where educators are currently to where they need to be. Utilize this tool thoughtfully—it's a critical addition to your coaching arsenal, serving as a foundational strategy for facilitating growth and unlocking potential at all levels within the educational sphere.

This checklist is a useful tool to help you facilitate the creation of highly effective IMPACT Goals (Figure 5.2). This resource is designed to be a practical guide, whether you are setting goals for an individual teacher's development, a grade-level team's progress, or for broader schoolwide objectives.

IMPACT Goals

INDIVIDUALIZED
The goal is relevant to the roles and responsibilities of the stakeholder(s).
The goal states specifically what the desired result is.
The goal is specific to the needs of the stakeholder(s).

MANAGEABLE	
	Staff have the knowledge and skills to reach this goal.
	Resources have been provided to support those who need to implement the change (e.g., time, tools, authority, etc.).
	Stakeholder(s) only focus on a few goals at a time.

POSITIVELY MOTIVATING	
	Stakeholder(s) understand why they should engage in the change.
	Stakeholder(s) are involved in developing the goal.
	Stakeholder(s) are motivated by the goal, or there is a plan for sustaining motivation.

ALIGNED	
	Specific behaviors and tasks required to achieve the goals have been identified.
	The behaviors and tasks identified are specific to the roles and responsibilities of the stakeholder(s).
	Achievement of this goal adds value to faculty, staff, and ultimately, the students.

CONNECTED	
	Accomplishments have been identified that can be used as salient measures of progress.
	The accomplishments are connected to the behaviors required to achieve them.
	The goal is connected to the accomplishments required to achieve it.

TRACKABLE	
	There is a process for measuing behavior required to achieve accomplishments.
	There is a process for measuring accomplishments required to achieve the goal.
	There is a process for tracking stakeholder perceptions.

Figure 5.2. *The IMPACT Goals Checklist ensures important components of goal setting are in place to increase the likelihood of goal achievement.*

Key Takeaways

- A school's culture, shaped by patterns of behavior, can be either a strong asset or a liability, and it's not set in stone. By applying the principles of ABA, leaders can systematically build desired behaviors, focusing on antecedents, behaviors, and consequences (ABCs), and understanding the pivotal role of consequences in sustaining and improving behaviors as crucial for achieving long-term success.

- Helping teachers succeed involves identifying desired teaching behaviors, recognizing that good teaching isn't based solely on personality but also on learned behaviors, and providing clear antecedents (training and instructions) to prompt those behaviors.

- It's important not to rely solely on antecedents, as sustaining positive behaviors requires understanding the role of consequences in the behavior change process.

- Consequences can be positive (something added) or negative (something subtracted), and they can be delivered by someone or they may occur naturally in the environment.

- Positive reinforcement fosters engagement, satisfaction, innovation, collaboration, and a supportive culture.

- Negative reinforcement might achieve short-term compliance but potentially strain the school's culture.

- While punishment can serve as a corrective tool, its misuse and overuse can lead to a negative environment, hinder skill development, foster resistance, and diminish motivation, emphasizing the importance of a balanced approach that prioritizes positive reinforcement for fostering growth and excellence.

- The key takeaway is that a balanced approach, with a strong emphasis on positive reinforcement and the careful use of corrective measures, creates an environment where individuals flourish and excel.

- Feedback is a versatile and powerful tool for improving performance, as it strengthens, refines, generalizes, discriminates, maintains, and chains behaviors while fostering self-evaluation, accountability, resilience, and a positive culture.

- Building habits for improved performance involves creating a strong connection between coaching suggestions and natural reinforcement. Effective coaching guides individuals toward accessing natural consequences that reinforce their desired behaviors.

- Goal setting involves collaboration and the creation of attainable goals that align with the individual's current performance level. Setting incremental subgoals and using reinforcement strategically can enhance the impact of goals on performance improvement.

- IMPACT Goals integrate principles and approaches into a structured framework, increasing the likelihood of achieving educational goals.

- While traditional SMART goals are useful, IMPACT Goals address additional critical variables for success.

- IMPACT Goals consist of individualized, manageable, positively motivating, aligned, connected, and trackable elements.

6

Building Teacher Performance

The importance of education, training, and professional development is most likely not lost on those who have dedicated their lives to educating others. Classroom and schoolwide results are driven by behavior, and these behaviors need to be developed. Students must be trained to meet the requirements of their curricula and their lives, teachers must be trained to teach the students, and coaches must be trained on how to coach the teachers. We've all been involved in training in one way or another.

Training can be a costly and time-consuming process, but scientific advancements can offer a path to take advantage of some of these innovative new solutions. Whether you are building performance online or through an on-site instructor, the exponential advancements in computer-based technology are enabling state-of-the-art learning environments that learners will both benefit from and gravitate toward. Computer-based technology is not the only area of scientific advancement that impacts training effectiveness—behavioral science accelerates learning and makes it more efficient.

Many seasoned leaders find that training and professional development can be the most frustrating and complex endeavors of any organization. Given the demands placed on leaders to evolve and produce results across different types of learners, regardless of barriers in their environment, it is easy to see how it might wear down even the most positive and energetic person. Whether you're an administrator, a new principal, a schoolteacher, or simply someone trying to learn a new hobby at home, training systems involve a number of critical pieces that must be respected in order to get to the ultimate critical goal: creating a skill that has demonstrable value.

Tackling these critical skills efficiently produces results, hopefully making life easier on both the leader and the direct report. But *are* these critical pieces respected?

Those wishing to become college professors are trained diligently for years in graduate school to be experts in their field. Once these individuals graduate, they will be tasked with taking their vast knowledge of physics, mathematics, art, or whatever else they dedicated a lifetime of student loan payments to, and teaching others to do what they do as educators. The problem? Just because you're an expert physicist doesn't mean you know how to teach people physics.

The issue isn't restricted to academia. Wayne Gretzky, who held over 60 National Hockey League records when he retired, had a winning percentage of under .500 as a coach. Isaiah Thomas, Ted Williams, Magic Johnson, and numerous other Hall of Fame athletes have shared a similar fate. Just because you're one of the greatest athletes in your sport doesn't mean that you know how to educate people in your sport. Just because you're good at math and science doesn't mean that you know how to educate people in math and science. And just because you were a good teacher doesn't mean that you know how to educate teachers.

Educating the Educators

The first step in an educator's education occurs when a person decides to become an educator. This journey can have a number of origins. Perhaps the individual comes from a family of educators and has always seen the value of helping others reach their potential. Maybe she was always really good with kids and saw teaching as a perfect avenue to inspire young people and build the next generation of enlightened and passionate citizens. Or maybe she had always been told that the way *she* explained something always made more sense. Teaching is a noble profession with endless demands and has been since the beginning of civilization. Once someone commits to becoming an educator, the next step is—you guessed it—to educate that person first.

But what do educators need to know? And how does that compare to the training teachers receive during their college and post-college education? For example, if you look at the average curriculum for instructional design degrees, you'll see numerous classes on learning systems design, assessment, computer technology—even a few on learning theories. But odds are you won't see classes on *behavior analysis*. This is unfortunate because behavior analysis, at its core, is the science behind learning as it helps us understand how actions, reactions, and responses shape our acquisition of knowledge and skills. Think about the classes you took as an educator. Were there more courses on subject content than on shaping behaviors associated with learning, getting students in touch with positive reinforcement, or perhaps other topics on the science behind dealing with behavior problems?

You most likely had more content-oriented classes. That's disappointing, as there's a science behind successfully building performance in and out of the classroom that goes beyond theory, beyond basics in subject matter and curriculum, and beyond relying too much on experience. Teachers who grasp the fundamental principles of behavior analysis are able to unlock insights into how students learn and respond. Understanding these basics empowers teachers to design effective strategies; cultivate positive classroom environments; and tailor instruction to meet individual needs, fostering optimal learning and growth.

Using Science to Teach

Each teacher brings their unique style, methods to engage students, and techniques to identify and use rewards. However, let's not forget the science behind skill-building and classroom management—it's crucial. Just as you wouldn't leave chemistry experiments to trial and error, teachers shouldn't have to figure out skill-building and classroom management through guesswork. But regardless of a teacher's style, there are some very specific skills that all educators should possess for managing behavior and creating an optimal learning environment (Latham, 1997). For example:

- the ability to teach expectations
- the ability to get and keep students on task
- the ability to maintain a high rate of positive teacher-to-pupil interactions
- the ability to respond noncoercively to inappropriate behavior that is consequential
- the ability to maintain a high rate of risk-free student response opportunities
- the ability to serve students with challenging behavior in the primary learning environment
- the ability to manage behavior scientifically

Successful teachers set up their respective learning environments so students can practice. Some of the activities are individualized (e.g., completing a worksheet, answering a question) and some of them occur as part of a group (e.g., working on a group project, responding in unison to a question), but they should all be targeted toward practice.

Practice makes perfect, right? Not quite. There is science behind what type of practice works. Ericsson et al. (1993) argued that practice must be "deliberate" to maximize performance, a claim that has been well-researched over the years (de Bruin et al., 2008). Deliberate practice centers on repetition and feedback that's both positive and constructive. The science has told us that students need significant opportunities for learning and practice (e.g., Johnson & Layng, 1992, 1994). We will reflect on this important concept again later in the book as it relates to Deliberate Coaching.

Looking through a scientific lens to understand why people do what they do can help build a classroom environment that maximizes teaching effectiveness. As we've discussed, one of the basic tenets of behavioral science is that behavior occurs because antecedents prompt the behavior and reinforcement keeps it going. A child raises her hand to answer a question because, in the past, this has been reinforced by the teacher praising the student. Generally speaking, the more opportunities for children to engage in these activities, the faster they'll learn. Add a speed component to your accuracy standards, and these skills can be learned even faster and last longer in the learner's repertoire.

This is one of the core teaching practices illustrated in the Morningside Academy approach. A learner has two options once a skill is taught: Use it or lose it. If the child uses the skill in day-to-day life or, at the very least, in subsequent lessons, then the child has continued opportunities to practice that skill and receive positive reinforcement for its use. If you teach a child to read a word, they have to keep practicing this skill when learning how to read sentences. If you teach a child to dribble a basketball, they have to keep practicing this skill when learning to play as part of a team. But if the child doesn't have the opportunity to practice these skills to fluency, they'll eventually go away.

School curricula should be built to ensure that the skills kids learn are either built into their day-to-day lives (to access natural reinforcement) or built into subsequent curricula. For those skills that aren't, the best-case scenario is to teach them in a way that makes them last until an opportunity arises to use them again. For example, using flashcards with spelling, multiplication tables, and foreign languages increases the odds that these skills will stick even if the child doesn't use them all the time. This type of fluency training (utilizing speed plus accuracy) has been shown to result in faster, longer-lasting skill acquisition than traditional training that only focuses on how accurate people are during a learning activity (Bucklin et al., 2000; Fox & Ghezzi, 2003).

Classrooms should be built to maximize learning opportunities. If teachers did not receive sufficient education on the science of behavior in college, then school coaches can step in to help. We need to ensure that teachers have all the tools they need to be successful!

USING SCIENCE TO TEACH TEACHERS

Let's meet two teachers, Ms. Rodriguez and Mr. Malek, who stand on opposite ends of the training spectrum when it comes to their classroom prowess.

In Ms. Rodriguez's class, her training shines through. Armed with a comprehensive understanding of behavior analysis principles, she seamlessly navigates her lessons. She anticipates student reactions and uses targeted prompts to guide their engagement. When challenges arise, she employs effective behavior management techniques, swiftly redirecting off-task behavior with minimal disruption. Her approach fosters a classroom culture of growth and mutual respect. Ms. Rodriguez's well-honed skills, acquired through extensive training, result in a harmonious and dynamic learning environment where students thrive.

Contrastingly, Mr. Malek, lacking in substantial training, faces hurdles in his classroom. While he's an enthusiastic educator, he struggles to decode the intricacies of behavior patterns. Often, classroom disruptions catch him off guard and his responses are inconsistent. His limited toolbox of behavior management strategies sometimes inadvertently escalates situations. Without the refined techniques that structured training offers, Mr. Malek finds it challenging to establish a consistent and productive classroom dynamic.

Through the experiences of Ms. Rodriguez and Mr. Malek, it becomes evident that effective teaching extends beyond enthusiasm. Training grounded in behavior analysis equips educators like Ms. Rodriguez with the tools to navigate the complex landscape of student behavior and to foster an environment conducive to learning. On the flip side, teachers like Mr. Malek, who lack such training, face an uphill battle as they struggle to maintain consistent control and to foster optimal learning outcomes.

It's clear—to bring out the best in our students, we must bring out the best in our teachers. But it's not possible, nor should it be possible, to get all of the pertinent information, experience, and practice needed to be a successful educator in college, even with the best college programs. Continued education and professional development are essential parts of building teacher performance that lasts. Such continued education and development can be proactive, giving teachers new and refined skills to set them up for success. They can also be reactive, undertaken as a result of a performance deficit. The ABCs of behavior analysis are an essential part of this process.

Effective training of critical teaching behaviors contains antecedents to prompt behavior in the training environment and consequences to build these behaviors. The type of behaviors learners engage in during training will depend on the type of training system. Traditional instructor-led training might involve learners taking notes, answering questions aloud during large-group discussion, or participating in group or individual exercises. Antecedents such as textbooks, training handouts, or lecture slides presented by the instructor are often used to get people engaged during the training. Instructor feedback, peer praise, and grades are examples of consequences that affect whether the behaviors being trained will continue and build. If we see more of these targeted learner behaviors as a result, then we say that *learning* has occurred.

Another form of training, computer-based instruction, offers a number of textual, auditory, and graphic displays as antecedents to prompt the types of behaviors possible on computer systems, including typing a response, dragging and dropping options, and clicking to select an answer. The science of behavior shows us that what happens during or after these behaviors will determine how fast people will learn. These consequences include events that take place within the program, such as textual praise acknowledging a correct answer, and those that take place outside of the computer-training environment, such as praise from an instructor or successful application of something learned.

Expertise in a subject does not automatically equate to the ability to teach it effectively, underscoring the need for specialized training in teaching methodologies.

Each kind of training program offers a unique way to build behaviors, and each learner has their own unique history of learning and their own preferred way of learning. But the building blocks of learning—the learning opportunities—are the same. A learning opportunity is the interaction of a training antecedent (e.g., an instructional statement) with a learner responding to that antecedent, and a consequence following that response (e.g., feedback). Decades of research have shown that learning occurs when people capitalize on the quantity and quality of these learning opportunities (e.g., Binder, 1996). Yet we are all familiar with training programs that present hours of lecture-based antecedents before a learner finally has an opportunity to respond, only for the opportunity to be in the form of a quiz that allows you to "pass" with 80% accuracy. You can't expect competent teacher or leader performance to occur when the only feedback provided occurs one time at the end of a long lecture in the form of some quiz. This is an error—a big one—that wastes time and money with no return on investment. Taking advantage of the science of behavior behind learning and performance can help maximize the impact of your training systems and avoid costly training errors.

If It Ain't Broke, Don't Fix It

Regardless of the information spanning years of research on what works and what doesn't work regarding training practices, the standard professional development day in schools often follows the same formula: Pick a topic or two, get a speaker to lecture about the topics, and hope that it sticks. But we know that doesn't work. The science behind educating educators—and more importantly, the vast store of tips and tricks we've learned from that science—is often overlooked or ignored. And these trends are going to continue. Why is this?

SOMETIMES IT WORKS

There are times when a trainer or a coach with no history of training or coaching can come out of nowhere and show you a unique skill set they happened to acquire during their previous work experience. You might just roll the dice and hope that this person does well, or you might have a strategic interviewing process that helps narrow things down. For example, if the person can describe the critical behaviors required of the job during the interview process, you'll have a certain level of confidence that they're not going to waste time with ineffective practices. You can compare the behaviors they cite with behaviors recommended in books such as this one to make sure they meet your standards. If a person cannot describe these behaviors, then you're really just crossing your fingers and hoping for the best.

Let's look at a couple of examples.

First, imagine a school district seeking a new curriculum coordinator. Among the applicants is Omar, a former chef known for his innovative culinary creations. In his interview, Omar not only exhibits a solid grasp of the critical behaviors needed for curriculum development but also shares his strategies for aligning learning objectives, assessing student progress, and fostering teacher collaboration. His experience in crafting intricate dishes has honed his ability to design intricate learning plans. The district's strategic interviewing process validates Omar's competency and demonstrates how skills from one field can transfer to education.

In contrast, consider a scenario where an elementary school is searching for a new math specialist. They come across Lisa, a former accountant and newly certified teacher, who expresses enthusiasm for teaching math. However, during the interview, Lisa struggles to outline the specific behaviors required for effective math instruction and student engagement. Despite her experience with math through accounting, the school realizes that Lisa's inability to articulate relevant teaching practices highlights a misalignment between her experiences and the role's expectations. In this instance, the strategic interviewing process becomes crucial to discern whether a candidate's unique background translates effectively to the education realm.

IT DOESN'T WORK—BUT YOU DON'T KNOW IT

Imagine you're the principal of a bustling elementary school. Your school has just implemented a new teaching approach aimed at improving student engagement and learning outcomes. However, despite the investment of time and money, initial excitement, and well-received training, you notice that something seems amiss. Students might be more active during lessons, but their test scores haven't shown the significant improvement you expected. It's time to reevaluate the success of your training systems, just as you would if you were in charge of developing professional development programs for your teachers.

Three key points come into focus:

- **Cost Isn't Always a Reliable Indicator:** While you invested resources into the new approach, you can't judge its effectiveness solely based on the expenses incurred. Just as an expensive training program might not guarantee improved learning outcomes, an investment in a new teaching method might not automatically translate into higher student achievement.

- **Test Scores Don't Tell the Whole Story:** Similarly, much as test scores or activity completion don't necessarily reflect students' ability to apply knowledge effectively, teachers' participation in workshops or passing quizzes doesn't always equate to successful implementation in the classroom. True success lies in the practical application of skills, not just the completion of assessments.

- **Participant Ratings Can Be Deceptive:** While high participant ratings might seem indicative of success, it's essential to question whether these ratings stem from genuine skill development or simply occur because the training was easy and unchallenging. High ratings can sometimes mask a lack of genuine progress.

Understanding these training metrics is crucial to making informed decisions about when and how to implement changes. But often, this precision can be overlooked. Without incorporating the science of behavior into instructional design and professional development, vital insights might be missed. It all comes back to behavior. For example, well-designed training includes the following components:

- well-defined learning objectives that outline what learners are expected to achieve by the end of the instruction

- content that captures learners' interest and maintains their engagement throughout the learning experience

- opportunities for learners to actively participate, practice, and apply the concepts being taught

- timely and meaningful feedback that guides learners' understanding and improvement

- connection between the content and real-life scenarios that allows learners to see the practical relevance of what they are learning

- a gradual progression from simple to more complex concepts to ensure effective learning and retention

Thankfully, there's a solution that includes all of the above and more for assessing and enhancing training systems so that learning generalizes into performance in the natural environment. This book unveils these strategies and offers practical methods to amplify the value of your training system. Regardless of your coaching or training platform, the toolbox provided can guide you toward preventing wasteful practices that could hinder your school's progress. By making subtle adjustments, you have the opportunity to embrace efficient methods for developing training systems and post-training coaching that are not only effective but will endure over time. Just as it is in your school, success is about more than superficial indicators—it's about nurturing genuine growth and advancement.

Key Takeaways

- Scientific advancements in technology and behavioral science can lead to more efficient and effective training methods, but it's important to recognize that expertise in a subject doesn't necessarily translate into expertise in teaching or coaching others in that subject.

- Understanding behavior through a scientific lens can help create an effective learning environment that maximizes teaching effectiveness and leverages principles of behavior analysis to ensure successful skill acquisition and classroom performance.

- Evaluating the success of new teaching approaches or training systems by relying solely on cost indicators, test scores, and participant ratings can be misleading. True success is measured by practical skill application, and incorporating behavior science into instructional design is essential for effective training that leads to enduring growth and advancement.

- Educational outcomes in both classrooms and schools are influenced by specific behaviors, necessitating their development among students, teachers, and coaches.

- Expertise in a subject does not automatically equate to the ability to teach it effectively, underscoring the need for specialized training in teaching methodologies.

- Understanding behavior through scientific principles of behavior analysis is key to enhancing classroom effectiveness, as behaviors are shaped by antecedents and reinforced over time.

- The effectiveness of different training methods, such as traditional and computer-based instruction, should be assessed based on how well they provide clear antecedents and reinforcing consequences to build effective teaching behaviors.

7

Misuse of Training

Most people spend more time and energy going around problems than in trying to solve them.

—COMMONLY ATTRIBUTED TO HENRY FORD

As the demand for training innovation increases, we can't leave behind the scientific foundation on which all learning is built. We can use this foundation to improve the way we educate the educators. We can build our training systems in a way that saves time and money without compromising quality and effectiveness. School districts have a prime opportunity to maximize the time spent on training and to maximize long-term impact if they will combine technological advancements with the evidence-based training methods discovered through years of research on the science of behavior. This is what the science of behavior has to offer: a set of tips and resources to ensure that your training expectations match your training development.

Even if schools are delivering their own training, it takes time, and time is money. A few errors here and there in the preparation and rollout of training can waste resources and cause undue frustration. For example, a subject-matter expert or senior teacher can present a seemingly endless list of competencies that need to be addressed in a given teacher-education program. It's tempting to teach all of these, and it's even more tempting to try to teach them all at once. But this is a mistake, and it can be a costly one.

Training priorities influence the use of resources. How these resources are used directly affects the return on investment that district and state leaders might be requesting. A clear understanding of training needs and optimal value provides an understanding of the specific behavior and results that will guide training efforts and set training systems up for success.

As a school leader trying to make the best use of time and resources, you should consider sifting through all the things that *can* be trained and narrowing them down to what *should* be trained. This leaves you with a set of behaviors that you're confident need to be trained; however, some of them will still be more critical than others. A trainer *can* spend professional development days teaching teachers and staff everything about the history of the school, advancements in school testing and why those occurred, different strategies they went through prior to deciding on implementing a Deliberate Coaching approach, etc., but there isn't time for that—and it isn't necessary for the teachers and staff to be successful. If you can't narrow the focus of your training to just those critical things that people need to thrive, then you're almost certainly setting yourself up for retraining. And if your training isn't effective in the first place, simply doing it over and over again won't magically make it work.

Every staff member in every school needs antecedents, including training, to prompt workplace behaviors and consequences such as positive reinforcement and feedback to keep those behaviors going. Coaches should make sure that the right antecedents and consequences are being used so the behaviors you want prevail and those you don't want decline.

Nonetheless, performance problems will persist. Things will go on that shouldn't, such as a teacher failing to develop rapport with students, setting low student expectations, engaging in inappropriate peer-to-peer gossip, or cutting corners. Or there might be behaviors you want that don't occur enough, aren't occurring at a high enough level, or aren't occurring at all. For example, say you want teachers to ask questions of their students. It becomes a problem if they're questioning the same students over and over again while ignoring the struggling students who are too afraid or unsure of their answers to immediately raise their hands.

So, again, the question becomes *why* are these problems occurring? Training is time-consuming and can be costly, and it's a waste of resources to apply it to a non-training problem.

The Quick Fix

Consider this example:

> A few years back, a school district hired a consultant to assist with their faculty performance management and evaluation programs. Part of the district's request was that the consultant guide the development of systems being integrated into a new school program designed to support children diagnosed with autism spectrum disorders. Prior to the consultant's involvement, there was an incident in gym class involving a disruptive child. The child was acting out, bothering the gym teacher and other students, and the situation escalated to the point where the

gym teacher pushed the child. The school requested training—specifically, training for all staff on general information about autism spectrum disorders and the ramifications of pushing any child.

Do you think the teacher believed that pushing a child was the right thing to do? Do you think a lack of understanding about autism caused the teacher to do something she knew was wrong? More importantly, do you think the proposed training would fix the problem and prevent further incidents like this one?

Therein lies the issue. We don't know. We can't know without asking the right questions, without assessing or at least having a reliable guess as to what caused the problem. Reliance on assumption, personal experience, and common sense can impede effective decision-making and will, more often than not, lead to errors.

The training systems in our schools make up just one of the many tools leaders have in their toolbox to help address performance problems. Training doesn't fix all performance problems, even if your professional development colleagues are consistently your first call whenever people make mistakes. The best leaders aren't just effective, they're also efficient. They know *why* a solution should be used—including training.

So why do we typically use training? The answer to this might seem obvious. Maybe all of the people receiving training are new teachers looking to complete their required onboarding process. It's impossible for each new teacher to start the first day of employment with a repertoire that's already filled with all of the critical school information. Maybe your school has new state-of-the-art technology that everyone needs to get up to speed with. The only way the technology will work is if people are taught how to use it, so training is a given. Or maybe there are slated professional development days, and training is what you do during these days. So, all of the teachers are sent to the same training, even though they may have varying needs.

But let's not forget the fundamental purpose of training: to build skills not yet in an individual's repertoire or refine skills in a way that produces maximum impact. When looking to achieve this goal in an efficient and effective manner, you can't only take into account how *well* training is delivered; what's driving the *demand* for training is equally critical. Training should be used to address a need for training. It shouldn't be a one-size-fits-all bandage for all performance problems. Understanding this is the best way to position you and your training services for success.

Using Training to Punish

In a crowded urban school, Principal Frick is deeply committed to maintaining high teaching standards. One day, he notices that Ms. Goulbourne, a dedicated English teacher, has been struggling with charting student progress. Concerned about this lapse, Principal Frick begins to consider how to address the situation effectively.

Understanding the behavioral dynamics at play, Principal Frick realizes that training can serve a dual purpose. While training is primarily aimed to build skills and prompt behavior, it can also be used to correct performance issues when necessary.

Principal Frick decides to take action. He believes that Ms. Goulbourne might lack a complete understanding of how to chart student progress accurately. With good intentions, he schedules a training session for the entire English department, assuming this will help address the problem.

However, unbeknownst to him, his well-intended plan carries some unintended consequences. The training, in this context, functions as a consequence, delivered in response to behavior. It inadvertently becomes a form of punishment for the inadequate progress charting, while simultaneously serving as an antecedent for the desired behavior targeted during the training.

As the training session approaches, teachers, including Ms. Goulbourne, feel the mounting pressure. The unintended punitive aspect of the training begins to overshadow its skill-building purpose. The atmosphere grows tense, and some teachers, including Ms. Goulbourne, feel discouraged and anxious.

After learning about Deliberate Coaching, Principal Frick soon realizes that while his approach aims to correct the performance issue, it has potential downsides. Improper use of training as a punitive measure can inadvertently diminish existing effective behaviors and negatively affect the overall climate and culture of the school.

With his newfound knowledge, Principal Frick recognizes the complexity of using training for corrective purposes; as a result, he seeks a more balanced approach. He decides to focus on addressing the root cause of Ms. Goulbourne's struggle through supportive coaching, individualized guidance, and collaborative problem-solving. By understanding the intricacies of using training as a tool for both building and correcting behaviors, Principal Frick aims to foster an environment that truly supports growth and improvement while maintaining a positive school culture.

Training is fundamentally designed to build a skill set and prompt behavior. Even with this underlying premise, however, it often has a dual purpose. When used as a management tool, training might be delivered *after* performance, particularly poor performance, as Principal Frick started to do in the story above.

As you now know, the science of behavior shows us that antecedents come before behavior to get it going, and consequences come during or after behavior to increase, maintain, or decrease it. Like Principal Frick, a principal might notice that a teacher isn't performing to a standard and assume that this is because the teacher doesn't understand how to do the specific task. And in the worst cases, the principal might not even question if the teacher knows how to do it. They might simply feel it's necessary to punish the poor performance and send a "message" to the teacher and others by requiring them to attend training. Regardless of intent, training in this context is being used as a consequence, delivered as a result of behavior. It's a punishment for the undesired behavior that preceded the training (e.g., teaching without charting student progress) in addition to being an antecedent for the desired behaviors targeted during the training (e.g., charting progress).

It's possible to use training to both build and to correct, as a proactive skill builder and a reactive performance management solution. However, if you use it improperly, you could inadvertently decrease existing behaviors that are effective instead of addressing the real cause of the performance problem. Moreover, if it's deliberately used as a punishment, the message you send may actually hamper the performance of the teacher, or even negatively impact the performance of other teachers as the result of the negative ripple effect that punishment can sometimes have on climate and culture.

Retraining for the Sake of Retraining

It's 7:15 a.m. at Greenfield High School, where a faculty meeting is in full swing. Principal Martin addresses the room, his concern evident as he discusses the persistent performance challenges faced by the teachers. As the conversations delve deeper, the topic of training emerges as a solution to address the lingering issues.

Principal Martin, believing in the potential of training, proposes that additional sessions be scheduled to enhance teachers' skills and knowledge. It seems logical—use training as a means to correct performance deficits and cultivate effective instruction. However, amid the optimism, a subtle but crucial concern looms.

Ms. Thompson, a seasoned teacher known for her insightful observations, raises her hand. With a measured tone, she questions the underlying assumption: "Are we certain that training is the sole solution? Could there be factors beyond lack of skills driving the performance issues?"

Principal Martin pauses, the weight of the question sinking in. The discussion shifts, and the staff begins to explore the possibility that not all performance problems are tied to a deficiency in skills. Maybe some stem from complexities within the classroom dynamics, or perhaps from external pressures influencing teacher effectiveness.

Ms. Thompson shares examples from her own experiences and highlights instances where training alone might not have resolved underlying challenges. She urges the importance of understanding the root causes before investing valuable time and resources into training initiatives.

As the conversation unfolds, it becomes evident that the assumption that training alone could rectify performance deficits is not always accurate. The realization that performance issues could stem from multifaceted sources and not solely from knowledge gaps marks a turning point in the meeting.

Principal Martin recognizes that the temptation to default to training as a universal solution is not uncommon, and the consequences can be costly. A misguided approach can lead to investing significant effort into training initiatives that fail to address the actual issues.

As the meeting concludes, a newfound awareness has taken root. The educators acknowledge that, while training remains an essential tool, it needs to be implemented thoughtfully and aligned with the true underlying causes of performance challenges. The team commits to a holistic approach, ensuring that the right interventions are applied to the right problems, ultimately leading to more effective solutions and improved teacher performance.

We've established that training applied as a reactive performance management solution is training applied in a dual role. If employed properly, this kind of training will target performance deficits. You teach someone because they have something they need to learn. This means that, based on either assessment or assumption, low or poor performance has resulted in a need for a solution, and you've identified training as the way to fix it. Training follows behavior as a consequence and is then used as an antecedent to build or refine the right behaviors.

At first glance, this seems to make sense. As part of its dual role, training is designed to get behavior going and, in the case of poor performance, get the desired behaviors going (e.g., effective instruction).

But there is a subtle yet significant problem here. Like Principal Martin, you're assuming that training will fix the problem, that a lack of training *is* the problem. But

Reliance on assumption can impede effective decision-making and will, more often than not, lead to errors.

what if that assumption is wrong? Are the performance deficits truly caused by a lack of knowledge and skills needed to do the job?

Always assuming that someone's performance problems occur because they lack a skill set can be demoralizing for the individual. In the vignette above, had Ms. Thompson not brought the issue to Principal Martin, what might have been the ramifications? How might it have impacted the teachers? The students? The school year?

We can attest that, regardless of industry and performance issue, training and retraining will be at the forefront in most discussions on performance-improvement solutions. You can most likely produce a series of examples. Maybe school leadership has been told that poor performance persists because there's something the professional development content missed. An administrator might propose that poor performance will be fixed if the teacher can only receive more training and see it modeled. Perhaps the teacher has been told that performance problems will go away if the teacher's assistants can only be trained on how important these duties are to the school. These misconceptions occur because we can't truly explain why we are training; we can't explain what caused the performance problem. Given the time, effort, and resources needed for training, this can be a costly error that leaves you with the same performance problems you started with.

Jack of All Trades, Master of None

It is the end of the school day at Middlefield Middle School. The leadership team is meeting, and the new principal, Principal Ramirez, takes center stage. With a purposeful tone, she discusses the upcoming training sessions for the school's staff. This training is designed to cover a range of topics within a limited timeframe. Besides it being warranted, there is also pressure coming from the state to build educator competency as the school has been targeted for improvement due to the previous year's poor performance.

The team members nod in agreement, cognizant of the challenges ahead. Among them is Mr. Foster, an experienced educator known for his insights into effective teaching strategies. Raising his hand, he steers the conversation toward a crucial consideration.

"As we plan these training sessions, let's keep the essence of learning in mind," Mr. Foster begins, sparking a meaningful discussion about prioritizing authentic learning experiences over mere administrative checkboxes.

He highlights the potential pitfalls of hurried training that might meet district or state requirements but fail to truly enrich the learning journey for both students and teachers.

"Middle schoolers are at a critical stage of development," he continues. "Quick training sessions might feel good because we're checking off boxes on our long to-do list, but they may not necessarily nurture the deep understanding required for teachers to create rich learning experiences for our students."

Mrs. Garcia, the technology coordinator, shares examples of rushed online modules resulting in superficial engagement among students.

Principal Ramirez listens thoughtfully, recognizing the balance needing to be struck between meeting training demands and fostering a genuine culture of learning within the school.

The conversation illuminates the need to ensure that training initiatives are thoughtfully designed to align with the school's mission. As the discussion unfolds, the leadership team reaches a consensus: Training should center on creating meaningful learning experiences for both students and educators. The team members agree that active engagement, opportunities for hands-on learning, and personalized feedback are essential components.

Principal Ramirez leaves the meeting with a renewed perspective. She understands that while administrative pressures are present, the school's commitment to fostering a holistic learning environment needs to guide the approach to training. The focus shifts from compliance to cultivating a vibrant culture of learning, where training initiatives contribute to the growth and development of the entire school community.

Even though training systems are ultimately designed to meet a performance need, there is no denying that there can be administrative pressures to get training done—and get it done quickly. As was the case for Principal Ramirez, it is very tempting to let this demand guide training development and delivery, resulting in training that's littered with way too many training targets and that, ultimately, doesn't work. Following this path most likely results in powerful outcomes including relieving administrative pressure or meeting some training quota. For Principal Ramirez, it probably would have felt great to check the training requirements off her long list of things that needed to get done.

But remember, if the training doesn't result in the trainees being able to perform the new skills well enough and long enough to produce some sort of valued outcome (e.g., increased student engagement and achievement), it isn't going to succeed! This leaves countless examples of workforce, college, and internet training modules where the desire to get everyone trained in a short period of time—regardless of how—is causing people to ignore the most important outcome. Misguided training is over-shadowing learning.

CHECKING A BOX VERSUS CHECKING FOR LEARNING

Once you've weeded out the behaviors that either aren't really important enough for training or are better suited for another type of antecedent, then it becomes a question of how much to train at once. A lecture or short presentation, followed by a quiz, will give you enough to claim that the training for the topic is complete. If this is the standard, it's relatively easy to cover a lot of training in a short amount of time—but there's a cost that will be obvious to you by this point.

This doesn't mean that lectures and short presentations are bad (they aren't), or that quizzes don't have a purpose (they do). What is missing is acknowledgement of the learner. Science tells us that learning requires responding. The more opportunities to offer responses and get feedback on them, the greater the learning. So training systems that are quick but offer limited opportunities for the trainee to perform and get feedback (i.e., that quiz you take at the end) will not produce learning. And if your training isn't producing learning, then the only thing it's really doing is checking a box for administrative purposes. Look at your primary outcome for training—it should have at least something to do with learning!

Organizing Training Content

The organization of training content should be based on four variables, all of which can affect the efficacy of the training and the reach of its impact.

1. SEARCH FOR CRITICALITY

Over at Riverside Middle School, Principal Thomas gathers his leadership team for a training discussion. The team includes various stakeholders, each with their own perspectives on training.

Principal Thomas starts the meeting by saying, "We have a lot of training targets to cover this year, including state mandates and new initiatives." He suggests compiling all these targets into a comprehensive training program. Mr. Bloom, an experienced teacher, raises a concern.

"Last year, we had a similar approach, and some teachers struggled to distinguish between what's essential and what's supplementary."

He recounts instances where teachers felt overwhelmed by the sheer volume of Information. Despite the concerns, the team decides to move forward with its scattershot approach, attempting to cover as many training targets as possible within a limited time frame. They do not prioritize specific skills or consider the varying needs of individual teachers.

On the other side of the district at Green Valley Middle School, Principal Hernandez sits down with her leadership team to discuss upcoming training plans. The team consists of dedicated educators and administrators who understand the importance of strategic training. Principal Hernandez opens the discussion by saying, "Our training content can come from various sources, but we need to ensure it aligns with our teachers' needs." She shares the story of a recent teacher survey that highlighted specific areas where improvement was needed. "Let's narrow down our training targets to address these critical skills," she suggests.

Ms. Turner, a veteran teacher, chimes in. "I've noticed that sometimes teachers feel overwhelmed by excessive information. If we focus on training what truly matters, we can maximize their engagement and retention." The team collectively agrees to prioritize a set of essential skills and develop training materials tailored to those areas. They discuss utilizing quick reference guides and job aids to reinforce these skills and support teachers in their daily tasks.

Principal Hernandez concludes the meeting by saying, "By homing in on the most vital skills, we'll create a more focused and impactful training experience. This approach will empower our teachers and ultimately enhance student success."

In the vignette above, both Principal Thomas and Principal Hernandez recognize the importance of training in optimizing teacher performance. However, their approaches differ significantly in terms of focus and strategy.

Principal Hernandez's approach exemplifies effective training focus. By identifying critical skills through teacher input and prioritizing those areas, the training content becomes targeted and meaningful. Quick reference guides and job aids are used to reinforce these skills, enhancing teacher engagement and retention.

On the other hand, Principal Thomas's scattershot approach fails to differentiate between critical and supplementary skills. The overwhelming volume of training targets results in confusion among teachers, potentially diluting the impact of the training.

In short, Principal Hernandez's approach demonstrates a more thoughtful and strategic way of training, whereas Principal Thomas's approach shows the potential pitfalls of trying to cover too much without prioritization. Effective training requires deliberate consideration of essential skills, ensuring they are not overshadowed by less critical content.

The content driving your training endeavors can come from a number of places. Maybe you've been handed a list of new mandatory education targets, or perhaps the content is standardized based on common requirements of the teachers in your state. If there's flexibility, consider whether these training targets actually *need* to be trained.

We know this might seem a little counterintuitive, given that it's the job of a school leader to ensure skills are built where skill deficits exist. But how valuable is the particular skill to day-to-day work activities? Much like Principal Thomas's approach, if you don't weed out the elements that don't matter, it can be difficult for teachers to tell the difference between the critical parts of the training and the parts they should be passingly familiar with. This can be extremely frustrating and cause them to under-value the entirety of the training.

If a particular target doesn't fit within your training priorities, don't worry; it doesn't mean it will be ignored. There are other ways to get the performance you're looking for. Prompts, including quick reference guides and job aids, are antecedents just as training is. Teachers have a lot of rules, names, procedures, acronyms, and so on that they supposedly need to know off the top of their heads. But do they really? Which ones are used most frequently? Which ones are better suited for a quick reference guide? Some of these rules and procedures might change so frequently that you'll spend more time rebuilding and retraining than you do with your initial onboarding. A well-designed checklist or reference guide can help employees perform accurately without wasting resources on teaching the hundreds of acronyms they might face on the job.

You need to give your training a fighting chance to build the necessary skills for an individual to sustain those skills over time and succeed. If a skill isn't going to be used often, there's no reinforcement in the environment to keep this behavior strong in the person's repertoire. It will eventually fade away.

Differentiating critical areas from those that are good to know but not integral to performance is an important step in organizing your training content. For instance, you might wish to train employees on how your school was formed, why it delivers the services that it does, and why things operate the way they do. But prioritizing things like *how* to deliver the service over things like *why* the service is delivered enables the employee to access more reinforcers and become more productive more quickly.

We have to make sure that we are adequately preparing teachers for the demands of the classroom. Just look at the stats in Figure 7.1. Our schools and our students can't afford to lose another teacher. The more teacher job openings there are, the more students are likely being taught by unqualified professionals or receiving subpar education as they are thrown into classrooms crammed with too many other students. In his book *The First 90 Days* (2013), Michael Watkins emphasizes the importance of capitalizing on the critical initial 3 months of employment. The faster you get a new hire on the job, happy, and productive, the faster you, and the employee, will see their value. New teachers and staff can have a large impact during this critical time period, and your training systems have a lot to do with it. Don't think for a second that just because they have a teacher's certification, they are prepared to meet all of the demands of the job within your particular school setting. Just as we want teachers to engage in formative assessments to guide instruction, school leaders must engage in formative assessments to guide teacher support based on their needs.

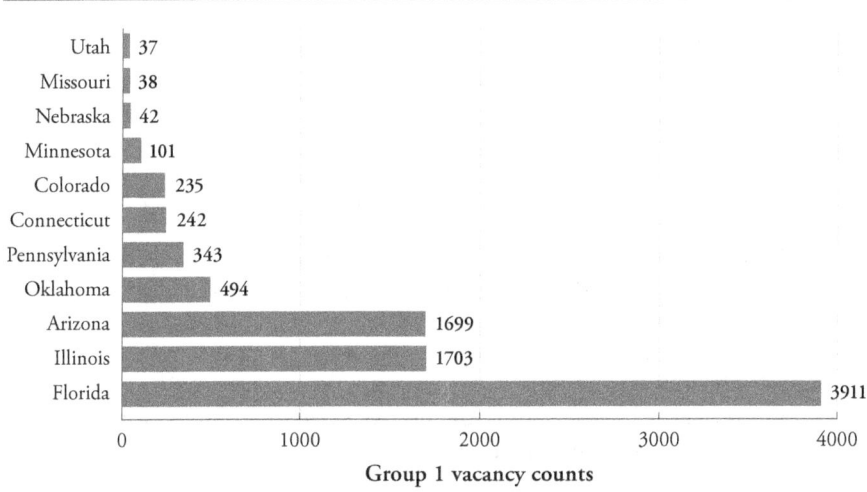

Figure 7.1. *Teacher job openings in public education directly impacts the quality of education students receive. Adapted from Nguyen et al. (2022).*

In some cases, the needs are associated with skill deficits, and the solution is well-designed and well-delivered training. In other cases, the teacher is suffering from a performance deficit. In these cases, they know what to do, but just aren't doing it for some reason, and they need coaching. So, understanding exactly what's at the root of a performance deficit is critical to determining whether somebody who isn't performing to a standard needs training or coaching. We'll cover performance diagnostics in a later chapter.

Your path to building good work habits starts by training the right behaviors and making the individual aware of the positive consequences associated with those behaviors. Many professions, including behavior analysis, will survey professionals in the field with the goal of producing essential training content. The results can lead to job analysis studies and other assessments geared toward finding critical areas that all practitioners in the field need to know. Prioritizing those critical behaviors in your training will save time and money and help maximize the value of your training.

2. BREAK DOWN COMPLEX SKILLS INTO SMALLER COMPONENT SKILLS

At Willowbrook Middle School, the faculty gathers for a professional development session on "Enhancing Student Engagement." The training, led by Principal Parry, aims to improve teacher skills in creating engaging classrooms.

Principal Parry begins, "Today, we'll dive into strategies for improving student engagement. Let's discuss the importance of building relationships, getting students engaged, and showing passion."

The teachers exchange puzzled glances. Ms. Bradley whispers to her colleague, "I'm not sure how to implement these vague ideas in my class."

Principal Parry proceeds with the training, providing a list of tips such as "use varied activities," "connect with students," and "be enthusiastic." These suggestions are presented as general guidelines, lacking specific details. As the session progresses, the teachers grow increasingly uncertain about the practicality of the training. The lack of concrete behaviors to implement leaves them feeling overwhelmed and unsure of how to apply the concepts in their classrooms.

Later, in the teacher's lounge, Ms. Bradley expresses her frustration, "I wanted actionable steps to engage my students, but the training felt abstract. I need more than just buzzwords."

The repercussions of the training's approach start to surface in classrooms. Teachers attempt to implement the vague concepts but struggle to translate them into actionable behaviors. As a result, student engagement does not show significant improvement. Principal Parry soon notices the disconnect between the training and classroom outcomes. Despite the intent to enhance student engagement, the lack of clear, trainable behaviors hindered teachers' ability to make effective changes.

In the example above, Principal Parry, though well-intentioned, wasted time and resources as she failed to break down complex skills into specific, actionable behaviors. This led to ineffective training that ultimately impacted student learning outcomes negatively. Just as you can save a great deal of time and resources by training the *right* behaviors, you can also maximize the efficiency of your training by looking at the *order* in which you're covering these behaviors.

By first identifying your targeted outcomes, you've set yourself up to separate the critical behavior pinpoints (remember, behaviors that are specific and measurable) you'll need to train to hit those targets.

> *The trainee's workday should be better after completing your training.*

General conversation about behavior often involves a focus on a descriptor or process such as "building relationships," "getting students engaged," or "showing passion." These are larger composite skills that are made up of the individual components (behaviors) that you've targeted for development in your training. The process of breaking down a complex skill into the component skills targeted for training is called a *task analysis*. Car mechanics initially need to know the key parts of the car, the function of each applicable mechanism, the steps to fixing various potential issues, and so on in order to effectively complete their duties. Teachers need to know what type of responding they want from students (e.g., choral, individualized, written), how to prompt these opportunities using well-planned instructions and questions, and how to respond to the students with feedback and reinforcement as they're participating in order to effectively build student engagement.

Training design should consider the experience level of the group and the complexity of the skill. If the group is less experienced or the skill is intricate, breaking down more complex composite skills into smaller components with clear steps and guidance becomes crucial for effective learning and application.

For example, breaking down the composite skill of "building relationships" into five component skills might look like this:

- **Active Listening:** Begin by teaching active listening as a foundational skill. This will set the tone for effective communication and understanding.

- **Individualized Attention:** Once active listening is established, introduce individualized attention. This shows students that you value them as individuals and are interested in their well-being.

- **Positive Reinforcement:** As the relationship deepens, incorporate positive reinforcement. This reinforces the positive interactions and behaviors between you and the students.

- **Shared Activities:** After establishing a positive rapport, engage in shared activities. This enhances your bond and provides opportunities for more natural interactions.

- **Behavior-Specific Feedback:** Finally, teach behavior-specific feedback. This allows you to highlight specific behaviors you want to encourage and helps create a positive environment for growth.

And still, the component skill of positive reinforcement can be broken down into smaller target behaviors.

For example:

- **Identifying Opportunities:** Recognize moments when positive reinforcement can be applied, such as when a student completes a task or demonstrates desired behavior.

- **Selecting Appropriate Reinforcers:** Choose reinforcers that are meaningful and motivating for the individual student to ensure they have a positive impact.

- **Timing Reinforcement and Its Proximity:** Deliver reinforcement immediately after the desired behavior occurs, and ensure it's directly linked to the behavior.

- **Using Verbal Praise and Approval:** Use specific and genuine verbal praise to acknowledge and highlight the behavior being reinforced.

- **Using Nonverbal Reinforcement:** Use gestures, smiles, or other nonverbal cues to convey approval and support.

- **Being Consistent and Predictable:** Establish a consistent pattern of reinforcement to reinforce behavior consistently over time.

- **Monitoring Progress:** Keep track of the effectiveness of the reinforcement strategy, and adjust it as needed for optimal results.

Even these can be broken down further if the trainee is unfamiliar with certain aspects. For example, progress monitoring might be further broken down into:

- Select appropriate assessment tools.
- Administer assessments consistently.
- Collect and record accurate data.
- Analyze data to identify trends.
- Adjust instruction based on data analysis.

Efficient training involves identifying critical behavior components and breaking down complex skills into measurable steps through task analysis. This approach maximizes learning outcomes and application effectiveness. When dealing with less experienced groups or intricate skills, breaking down composite skills into smaller behavioral components with clear steps becomes essential for successful training. Breaking down these skills will help refine your training and ensure that nothing is being missed.

3. FIND YOUR BEHAVIORAL CUSPS

In a busy middle school in a rural area, a trainer is tasked with improving teacher effectiveness. The trainer conducts training sessions on various aspects of classroom management and instructional strategies. However, the trainer doesn't consider the concept of teaching students how to self-monitor their progress as a training target.

While the teachers gain some insights, they struggle to see the practical impact of the training in their day-to-day work. The sessions lack a targeted approach that could open doors to new behaviors and opportunities. As a result, teachers continue to face challenges in their classrooms, and the training does not lead to the desired improvements in student learning and teacher satisfaction. The students' progress remains somewhat stagnant, as they lack the tools to take ownership of their learning.

Across town, another trainer takes a different approach to training middle school teachers. Recognizing the power of teaching students how to self-monitor their progress, the trainer strategically focuses on this skill during training sessions. This not only empowers teachers but also introduces a valuable behavioral cusp into the students' learning journey.

By enabling teachers to teach students how to assess their understanding, set goals, and track their growth, the trainer transforms the classroom dynamics. As students learn to self-monitor, they gain not only academic skills but also valuable life skills. This shift in focus leads to improved student engagement, better communication, and a more positive learning environment. Moreover, the concept of self-monitoring generalizes into other areas of learning, empowering students to take charge of their education and become more independent learners.

The trainer who ignores incorporating the "teaching students how to self-monitor their progress" concept delivers generic training that leaves teachers struggling to apply the concepts in their classrooms. On the other hand, the trainer who embraces this specific behavioral cusp strategically selects a training target that not only improves teacher effectiveness but also empowers students to become proactive learners.

The latter approach demonstrates the importance of considering the potential impact of training on various levels—individual, classroom, organization, and community—ultimately leading to more effective teaching, improved student outcomes, and a broader impact on student life skills.

The component skills identified through your task analysis will help direct your training efforts and will build both the individual and the more complex composite skills. If we dig around in our science toolbox, there is another behavioral concept that helps select training targets. These targets take things a step further, opening doors to *new* behaviors and *new* opportunities, exactly as the trainer did in the story above when she trained teachers in teaching students how to self-monitor. These types of behaviors, like self-monitoring, are referred to as *behavioral cusps*. Other behavioral cusps for teachers might include behaviors such as teaching effective classroom routines, implementing formative assessment strategies, or promoting active student participation through opportunities to respond. In each case, these approaches have broader positive implications. Let's dig into the concept of behavioral cusps a little more.

In 2001, a pair of researchers from Western Michigan University published a model for selecting target behaviors based on the concept of behavioral cusps (Bosch & Fuqua). The more criteria a behavior meets, the stronger it is as a training target, and the more you and the performer will experience a return on investment.

- **Does the skill help the person access reinforcers and desired environments?** We learn letters, numbers, words, and other reading components in order to engage in reading activities throughout our lives. Reading is an important behavioral cusp in that it helps us access other reinforcers that will in turn help build new behaviors. Reading means you can now access fun books and magazines, and eventually even that book on Deliberate Coaching that will help you in your new vice principal or instructional coaching job!

- **Does the skill help other people?** Training proper communication strategies is an important behavioral cusp that has a significant impact on all parties. Teachers and coaches in education have a broad impact on those around them. A teacher learning how to properly communicate student concerns with parents can help them get the type of parent support needed for each student to succeed while also helping the student and peers working with the student. The parent will see how communicating and adhering to teacher instructions helps their child and, in turn, can start seeing the value of working more with the teacher.

- **Does the skill benefit the organization or community?** Of the many behaviors targeted during leadership training, the ability to provide precise, sincere, and consistent positive reinforcement is the most critical. Why? It's not only good for the leader providing the reinforcement (leading to improved relationships) and the direct reports receiving the reinforcement (who learn faster and develop improved self-esteem), but it adds positivity and growth to any system. The better the skill fits within the classroom, the more likely the skill is to have sustained value that transfers beyond the classroom doors.

- **Does the skill help generate other desired behaviors?** Schools use code systems to publicly communicate various messages to staff. Knowing and reacting appropriately to these codes is an important behavioral cusp because it has a range of uses. It can help you avoid dangerous situations and protect students, it can tell you where to go and when you're needed, and it can show you what to avoid so as not to make things worse. Along with benefiting the staff and students, learning these codes can generate new behaviors (what to do when you hear the code), new opportunities for reinforcement (for following the code), and new situational rules (what happens in code-related situations).

- **Does the new skill replace an ineffective, inefficient, or otherwise inappropriate skill?** If your training is to serve as a solution to a performance problem, your behavioral cusp could be a new, more desired skill. Perhaps teachers are following the wrong procedure, cutting corners that result in costly or dangerous errors that negatively impact student achievement, or are using an outdated method that takes too much time and yields fewer results. Training new skills to replace these costly skills can add significant value.

"The difference between ordinary and extraordinary is practice."

—COMMONLY ATTRIBUTED TO VLADIMIR HOROWITZ

Train people how to ask for a fish when hungry, and they'll be able to ask for and maybe even receive a fish. Focus on a behavioral cusp, and they'll know how to catch their own fish and perhaps feed other people if necessary.

4. FOCUS ON FLUENCY

So you've separated your need-to-know skills from your good-to-know skills and prioritized the order in which to train them. If they're critical, then you want expert-level responses that result from your training. You want them responding fluently (quickly and accurately).

Say you're a beginning guitar player learning a new song. You wish to wow your friends and family at the local holiday party by showing off your improved chops. You commit, get people excited to see what you've been learning, and maybe even brag a bit because you've made some fast progress on learning your fingerings. As you start to practice, however, your nerves turn into intense fear as you realize that you don't yet have the tools to play a whole song, and your performance might be a colossal embarrassment.

A song is a combination of a number of component skills—identifying different chord structures, forming chords on the fretboard, moving from one chord to another (quickly), strumming in rhythm—and doing it all so seamlessly that you can lead a group in an off-pitch rendition of "Jingle Bells" without skipping a beat. So you spend every waking hour deliberately practicing each component skill one by one, then in combination, until you can do it. You become accurate and you become fast—you become fluent enough to take your guitar and ugly Christmas sweater to the party with confidence.

This same fluency approach taken when learning component skills applies to effective workplace training systems. But you might be thinking, "playing a guitar is one thing, teaching is another!" And while this is behaviorally correct, the principles behind training and fluency development are exactly the same. Let's take a look at Principal Tran's and Principal Davis's approaches to presenting a training on delivering positive reinforcement.

> In a group professional development session facilitated by Principal Davis, a team of teachers participates in a nonfluency-based training approach for delivering positive reinforcement. The session primarily comprises theoretical lectures and limited modeling of reinforcement concepts. Although Principal Davis is able to get through a number of slides in his PowerPoint and cover a lot of content, the lack of targeted practice and repetition hinders the teachers' ability to confidently apply reinforcement techniques.

> In an alternative school, Principal Tran adopts a fluency-based training approach to instruct a team of teachers in delivering positive reinforcement. While she provides some lecture, the bulk of the training is made up of a number of interactive role-playing exercises where the teachers engage in practicing specific verbal praise and tangible reward delivery. They collectively analyze performance scenarios, identify instances of effective reinforcement, and discuss strategies for improvement. Principal Tran provides real-time feedback to the group and ensures repeated practice to facilitate seamless integration of positive reinforcement techniques into their classroom management strategies.

In the scenarios above, Principal Davis may have covered more through his lecture style, but it was Principal Tran's approach that included practice and feedback using different approaches that won the day.

One valuable lesson you might consider when developing training at your school is that if you teach component skills to fluency, more complex composite skills will occur more quickly and easily. If you can spell and sound out words fluently, you're more likely to pick up reading and writing. If you're fluent in the parts of a cell, you're more likely to excel when teaching students these parts and troubleshooting student barriers. As you design your training, the type of responses you'll require from your participants during training (e.g., fluency versus mastery) can dictate the success of your training.

Behavioral Skills Training

In this chapter you've learned about the importance of identifying critical skills, breaking them down into components, identifying the all-important behavioral cusps, and training to fluency. Now let's wrap it all up into a tight little package where we demonstrate what you've learned so far through actional steps. Fortunately, there is a simple yet well-researched process for developing skills that incorporates many of these strategies. This process is called behavioral skills training (BST). In short, it employs instruction, modeling, rehearsal, and feedback as a system for developing a variety of performance skills. Reid and Parsons (2006) suggest the following steps:

1. Specify skills to be taught (declarative knowledge).

2. Provide training participants with a written summary of skills specified (declarative knowledge).

3. Verbally describe the skills with a rationale (the why!) for targeting the skills for training (declarative knowledge).

4. Physically demonstrate the skills for the participants (declarative knowledge).

5. Observe the participants practice the targeted skills (procedural knowledge).

6. Provide corrective or supportive feedback to the participants based on their demonstration of the skills (procedural knowledge).

7. Continue steps 4, 5, and 6 until the participants can perform the targeted skills correctly (procedural knowledge).

At this point you might be thinking, "How the heck can we give a training where participants get to practice *and* receive a steady stream of feedback?" This is a good question to consider. Classroom models of training are great for accommodating a large group of people but are often poor at providing them with opportunities for adequate practice and feedback. The solution: peer feedback.

As you know, each skill is a chain of behaviors that leads to a desired outcome. This chain can be broken down into a set of smaller written steps or components, as we outlined in the task analysis. This is exemplified by Reid and Parsons's seven steps for developing skills outlined above. Once these steps are listed, they can be used by participants to guide and provide feedback to one another as a cooperative learning strategy.

> *Classroom models of training are great for accommodating a large group of people but are often poor at providing them with opportunities for adequate practice and feedback. The solution: peer feedback.*

Let's unpack this a bit further to demonstrate task analysis as an extremely valuable tool. In fact, this tool might be considered a behavioral cusp for both teachers and coaches as it can be used to achieve meaningful outcomes under a variety of conditions.

Imagine a trainer is teaching participants to use *behavior-specific praise* to improve academic achievement. For example, "Johnny, you got the math problem right because you remembered to carry over the 1. Nice job!"

Or perhaps the focus is on improving behavior, "Johnny, thanks for raising your hand and waiting to be called on before speaking."

Behavior-specific praise is a skill that can be broken down into a series of components and steps. Once these are broken down into steps, they can be written on a sheet of paper as a checklist and used by participants to score one another regarding their declarative knowledge (i.e., Can they state the steps?) and their procedural knowledge (i.e., Can they perform the steps?) during practices.

There are many cooperative learning strategies for increasing feedback and engaging learners during instructor-led training models. Kagan and Kagan (2009) list a variety of these strategies that can be employed with both students and educators. Figure 7.2 shows how the process of a task analysis might look using what Kagan and Kagan call the Inside-Outside Circle.

Performance Feedback Task Activity: **Describing Key Elements of the Skill**

Description: Teachers rotate in concentric circles to face new partners for sharing, quizzing, or problem-solving. Teachers use the Describing Skill performance task sheet to interview each other in the Inside-Outside Circle.	
Setup: Presenter provides the definitions, role-plays, and allows the teachers to score the presenter using the performance feedback sheet in the P column.	
Step 1	Participants are asked to stand up with their Describing Skill sheet and form pairs. One teacher from each pair moves to form one large circle facing outward. The remaining teachers find and face their partners (class now stands in two concentric circles).
Step 2	Teacher exchanges their Describing Skill sheet with their partner, then asks their partner to describe the skills without looking at the sheet. Participants score each other and return their Describing Skill sheets to each other.
Step 3	Trainer directs inside-circle participants to rotate clockwise to a new partner. Trainer may call rotation numbers (e.g., "Rotate three ahead").
Step 4	Presenter asks participants to find another partner and repeat until the Describing Skill sheet is completed and participants have reached established performance goals.

Figure 7.2. Task analysis to illustrate the breakdown and chaining of components for teaching participants to accurately describe performance feedback using Kagan and Kagan's (2009) Inside-Outside Circle.

This process provides a safe way for participants to practice and receive precise feedback through a task-analyzed checklist. This checklist can include a question about the why, as illustrated in the Describing Behavior-Specific Praise (a behavioral cusp) table (Figure 7.3). While this example only lists six intervals beyond the initial P (presenter) column, the number can be increased based on the complexity of the task to allow for more repetitions and feedback. The goal might be for the participant to accurately describe the steps three times in a row before moving on to the next step of actually performing the skill.

P		1	2	3	4	5	6	Describing Behavior-Specific Praise
Y								Teacher should make positive comment about specific behavior immediately following the behavior.
N								
Y								Teacher should align the behavior with progress toward a goal.
N								
Y								Teacher should appear sincere (smile!).
N								
Y								Effectively used when teachers are trying to increase or strengthen a desired behavior.
N								

Figure 7.3. Task analysis to illustrate the breakdown and chaining of components associated with a teacher being able to accurately describe behavior-specific praise.

Once the teacher can describe the skill accurately, they should be given the opportunity to practice it using the same process. The Performance Feedback Task Activity table (Figure 7.4) shows what that process might look like. Remember, the previous Describing Behavior-Specific Praise activity requires the performer to *describe* the skills, whereas the process that follows it requires the performer to actually *demonstrate* the skill while the peer checks off which components are implemented correctly and which are implemented incorrectly. By engaging peers in the process of measurement and feedback, performance across the entire group is accelerated.

Figure 7.5 shows an example of the corresponding performance checklist. This checklist is intended to measure performance or application of procedural knowledge as opposed to describing the skill or declarative knowledge.

In another example, the trainer teaches participants how to use an attention signal. Figure 7.6 shows what that task-analyzed performance sheet might look like.

	Performance Feedback Task Activity: **Performing Key Elements of the Skill**
Description: Teachers rotate in concentric circles to face new partners for sharing, quizzing, or problem-solving. Teachers use the Performing Skill performance task sheet to score and provide feedback to each other in the Inside-Outside Circle.	
Setup: Presenter provides the definitions, role-plays, and allows the participants to score the presenter using the performance feedback sheet in the P column.	
Step 1	Participants are asked to stand up with their Performing Skill sheet and form pairs. One teacher from each pair moves to form one large circle facing outward. The remaining teachers find and face their partners (class now stands in two concentric circles).
Step 2	Teachers exchange their Performing Skill sheet with their partner. The trainer provides participants with a scenario highlighting a student behaving in a way that a teacher would like to see repeated.
Step 3	The participant asks their partner to demonstrate the skills based on the scenario provided. Without looking at the sheet, participants score each other, then return their Performing Skill sheets to each other. Trainer directs inside-circle participants to rotate clockwise to a new partner. Trainer may call rotation numbers (e.g., "Rotate three ahead").
Step 4	Presenter asks participants to find another partner and repeat until the Performing Skill sheet is completed based on an established goal (e.g., performing each step correctly three times in a row).

Figure 7.4. Task analysis to illustrate the breakdown and chaining of components for teaching participants to accurately be able to engage in performance feedback using Kagan and Kagan's (2009) Inside-Outside Circle.

Practice and feedback are where abstraction meets application in training focused on skill development. We're probably starting to sound like a broken record here, but that's okay. Without practice and feedback, training is a waste of precious time and money. Using behavioral skills training and engaging participants in performance-based peer feedback as illustrated in the previous example is a powerful approach to ensuring that participants actually leave the training with the ability to apply skills within the school.

P	1	2	3	4	5	6	Performing Behavior-Specific Praise
Y							Teacher made positive comment about specific behavior immediately following the behavior.
N							
Y							Teacher aligned the behavior with progress toward a goal.
N							
Y							Teacher smiled.
N							
Y							Teacher explained it should be used when attempting to increase or strengthen a desired behavior.
N							

Figure 7.5. Task analysis to illustrate the breakdown and chaining of components associated with a teacher being able to accurately perform behavior-specific praise.

P	1	2	3	4	5	6	Attention Signal
Y							Teacher used an auditory prompt (e.g., "Your attention please").
N							
Y							Teacher used a visual prompt (e.g., hand raised).
N							
Y							Teacher waited for students to stop, look, and listen before providing direction.
N							
Y							Teacher explained the attention signal should be used to gain student attention prior to providing a new directive to the class.
N							

Figure 7.6. Task analysis to illustrate the breakdown and chaining of components associated with a teacher being able to accurately perform the attention signal.

CONDITIONAL KNOWLEDGE

In business, good executive coaches understand that talent development involves a process of instructing, modeling, and observing paired with constant feedback to shape the skill. This is the development of the declarative and procedural knowledge discussed in Chapter 2. These coaches also understand that transferring skills learned in practice to actual performance in the organization is critical to success. Some leaders who make amazing speeches and drop inspirational quotes at the annual staff meeting will regularly demonstrate critical errors in leadership that impact their bottom line—they talk the talk but fail to walk the walk.

Sometimes poor performance can be linked to failure of skills to generalize from one environment to another, as conditional knowledge wasn't nurtured during the talent-development process. People know the why and the how, but they don't behave the right way at the right time.

Conditional knowledge is developed experientially; it allows the performer to understand the "when" while strengthening the "why" and "how" in the context of the work environment. It connects the skills and knowledge learned through deliberate practice and allows the learner to apply this knowledge within various aspects of the job. In teaching or school leadership, conditional knowledge allows the educators to follow school improvement plans as they apply the right technique at the right time. From a behaviorist point of view, this knowledge can be represented across multiple people and tasks through a series of "if-thens." You remember, those metacontingencies discussed in Chapter 1: "If this happens, then I will do that."

For example, much like Principal Tran, in various businesses, good talent developers often use performance-based scenarios or *simulations* to allow their employees the opportunity to safely apply learned skills under conditions that closely mimic the actual work environment. In fact, you see this all of the time in sports. For example, in boxing, a simulation is referred to as sparring. These fight-like conditions are critical for preparing the boxer for the real-world event—the fight!

Under these conditions, performance can be assessed and shaped by the trainers as the performers engage in actual activities required to be successful in the work (school) environment. Simulations offer a powerful strategy for increasing the likelihood that performers will be effective on the job.

Business mogul Richard Branson (2014) was quoted as saying, "Train people well enough so they can leave, treat them well enough so they don't want to." If districts and schools wish to help leaders and educators reach their greatest potential, teacher preparation and training programs must focus more on helping them perform to their potential by providing effective talent development that requires rehearsal under "fight-like" conditions. Training is absolutely critical for equipping educators with the skills they need to be successful. But talent development doesn't end in the training sessions. An effective approach must help the educator to transfer the skills they learn into the actual school—a Deliberate Coaching process.

Use It Wisely

Training, in all its forms and functions, is first and foremost an antecedent. It is used to set people up to succeed and provides each trainee with a repertoire critical to their job function. It can both *motivate* and *guide* learners.

Reflect back on the concept of using training to address performance problems. If there is a true need for training, then you should be able to link the training to something meaningful to the individual or school, something that makes it worthwhile and fixes the problem. If learning is fun, interesting, and presented in a way that demonstrates value, then the trainee will want to learn. This is a critical feature of any skill-building system, be it training, coaching, or some related initiative. It builds exemplars with strong productive work habits and helps create an organizational culture of success. You want people seeking out your training because they've found it adds value and fixes something they see as broken.

With typical training, the burden of learning is unjustly placed on the individual learner. They should participate in training, regardless of why it's being administered, because it's their job and that's what they're paid for. But while job requirements will get them in the door, the training is doomed to fail if that's the only reason they think they're participating in training. If motivation during training is low, then you shouldn't expect the motivation to *use* whatever was trained to be much higher.

If you want teachers to apply your training properly, then you need to show objective impact, particularly impact during those times it's being used in the way that you want it to be used. What does the teacher need to succeed in a positive and efficient way? And how will the training help the teacher do this?

Maybe a teacher has struggled to manage some disruptive students, which impacts the progress of those students and others in the classroom. If you can link the training you're providing with the progress the teacher is making in this area, then you're showing measurable impact—you're showing value. Without measurable impact after training, the training will appear to be useless, and if training isn't the solution (that is, if the individual can already demonstrate everything being trained), then the training may very well *be* useless. Why do it if the training adds nothing of note to the individual or the classroom?

The burden of learning is on the trainer, those building the training, and those making decisions about how to use the training. Those are the individuals responsible for motivation, and motivation is what drives learning and subsequent performance.

School training systems should be given a chance to add value to the teachers and staff participating in them. If not, the training days at your school, and those working hard to build and deliver those trainings, risk losing value. Here are a few things to consider when looking to maximize the impact of training services on performance problems.

1. RULE OUT ENVIRONMENTAL BARRIERS

As part of a 2005 study on performance assessment, store managers at a chain of sandwich shops stated that one of their top priorities was to get their staff to offer promotional stamps when a customer purchased a sandwich (Rodriguez et al., 2006). This was important to the stores as it helped drive future business, but the staff was struggling. A key discovery made during the study: The machine that dispensed the promotional stamps was broken.

When you look at the many organizational assessments researched and utilized over the years, one recommendation remains clear: Don't be quick to assume it's the performer's fault. Malfunctioning equipment, vague or nonexistent policies and procedures, and other systems barriers can impact performance as much or more than a lack of knowledge and a lack of consequences. If the teacher knows how to run a program but there are no materials for that program in the classroom, no amount of training will fix this problem. Rule this out first.

2. FIND THE TRUE CAUSE OF THE PERFORMANCE PROBLEM

You've ruled out environmental or process barriers hindering performance, so now what? Now we get back to our ABCs. When deciding whether training is the answer to a performance problem, see if the individual can demonstrate the task and demonstrate it at a high level. If so, it's not a knowledge or skills issue. Because the fundamental purpose of training is to address knowledge and skill deficits, spending time on training in this case would produce little to no benefit. You'd be better off using other antecedents, such as reminders or checklists, and providing effective and consistent feedback. If the individual cannot demonstrate the skill, then your training can be used as a reactive solution, specifically targeting the missing skill set.

3. IDENTIFY YOUR PROACTIVE-TRAINING NEEDS

When used as a proactive skill builder as opposed to a reactive one, training systems target a seemingly blank canvas. Again, based on either assessment or assumption, training provides knowledge and skills that individuals either don't have in their repertoire or don't have at the target level. The goal here is to help people understand what the leaders and training instructor deem integral to success given a new role or new change initiative. This might be directed toward using a new process or piece of equipment, getting new hires up to speed on critical procedures, seeing that key individuals know how to fix an impending problem, or another area of skill-building priority. The purpose behind proactive training is to provide a repertoire that the trainees don't already possess. The key for trainers is to properly identify that repertoire and relay the value of this skill set during the training.

One Last Point on Trial and Error

As you look around your school and your life outside of work, you'll find there is no shortage of opportunities to use training to address performance deficits. Assessing the cause of the performance problem can save you a lot of time, money, and frustration. However, you might still be on the fence about whether it really is a problem to rely too much on trial and error in your selection of performance solutions. Yes, moving toward a strategic use of training and other solutions means moving away from a proven and familiar methodology. Yes, it takes time to accommodate the trials, and there is a cost to the errors. Nonetheless, the solution might still appear at the end of this path. So, what's the harm in guessing?

Three outcomes can result from discarding your scientific know-how, none of which provide optimal long-term solutions from a leadership perspective.

- *Outcome 1—You guess right.* This is your best-case scenario. You find that training happened to work and you do the same thing the next time, hoping that you guess right again. But if your guess is wrong the second time and it's not a knowledge or skill issue, you'll be back to square one. Good leaders know why their leadership initiatives worked and can replicate results.

- *Outcome 2—You make it worse.* Suppose you have a select group of highly skilled, well-trained teachers who are working hard to hit their student-progress measures despite faulty equipment, missing materials, absent teacher assistants, and other issues out of their control. Individually, they're going above and beyond. However, lower performers are bringing down the success of the school as a whole, and as a result, you force the entire group to attend retraining on the basics of teaching standards and classroom goals. By assuming a knowledge or skill deficit of the entire group, you've punished the high performance of your exemplars. The end result is a group of low-performing teachers, still performing below standard after the needless retraining, as well as exemplars who feel ignored and disrespected, leading to declining performance across the board.

- *Outcome 3—Nothing happens.* You might not have fixed the problem, but you also didn't make it worse. You lost some time and money, and your teachers and staff returned from training to the same situation they were in before.

A FRAMEWORK FOR PERFORMANCE IMPROVEMENT

The science of behavior offers a level of precision that seeks out the causes of behavior in order to maximize the effective leadership tools you have in your repertoire, including the proper use of valuable training and professional or talent development systems. These are evidence-based tools to help you assess, manage, and sustain any issues related to learning and performance. A key discovery in this science is that behavior occurs or doesn't occur for a reason. This reason, whatever it is, should guide performance-improvement initiatives such as training.

Without a reliable understanding of *why* a particular performance problem has occurred, you are lacking the most critical piece of information—information that should be guiding not just your training but all your leadership efforts. Without knowing the cause of the performance problem, you're losing many of the advantages of a scientific analysis of learning and performance. By skipping the sound practice that has been proven to produce faster and longer-lasting performance improvements for your school personnel, you're not allowing yourself to truly maximize training and other leadership solutions.

We're offering a consistent and replicable framework. New behaviors are trained by presenting the right instructions, training materials, and other antecedents to get these skills going and using positive reinforcement to accelerate the learning, effectively building these skills into a person's repertoire. Select these training targets carefully. If you do this, you can maximize the time spent training and provide real value to the learner. The trainee's workday should be better after completing your training. Can they now do a job they couldn't do before? Can they do it more productively? Is their job easier? Can they and others see the value in these newly trained skills? You can find great value in your training via a quick examination of how you're prioritizing critical training content.

So how should you use the training systems you and your team have worked so hard to build? Give learners the tools they need to be successful. Take into account that the ABCs of our science require antecedents, and as trainers, we're responsible for one of the most important of those. And perhaps most importantly, make sure the problem justifies the use of training.

Training is a valuable necessity in every person's life. As leaders, we just need to capture and maximize that value by appreciating how and when training is used. If a coach uses tools such as training and coaching properly, and people see the value of the coaching, then more people will want the coach around and will benefit from what the coach has to offer.

Key Takeaways

- Training should be used strategically to address specific needs and build or refine skills, rather than being a generalized solution for all performance problems. Understanding the underlying causes of issues is crucial for effective decision-making in training.

- Training should be used to build skills and prompt behavior, but it can also inadvertently function as a consequence if used to correct poor performance. Understanding the distinction between using training as a proactive skill builder versus a reactive performance management solution is crucial to maintaining a positive school culture and effectively addressing performance issues.

- Relying solely on training as a reactive solution to performance deficits assumes that lack of knowledge and skills is the root cause of the problem.

- There might be administrative pressures to complete training quickly. Focusing solely on checking off boxes and meeting quotas can lead to ineffective training. The real value of training lies in its ability to facilitate learning and improve skills, not just in meeting administrative demands.

- Not all skills or information needs to be trained equally. Focusing on the most essential skills for day-to-day work activities can lead to more effective training outcomes.

- By breaking down complex skills into manageable components and providing clear guidance, training can be tailored to the experience level of the participants, leading to more successful learning outcomes and improved job performance.

- By selecting and training behavioral cusps, you not only improve individual performance but also create ripple effects that enhance overall outcomes and contribute to growth and positive change within the organization or community.

- Training should focus on practicing component skills to fluency, as this leads to more confident and successful application of complex composite skills in various domains, including classroom management and teaching strategies.

- Employing a structured approach like BST, which involves instruction, modeling, rehearsal, and feedback, can effectively develop performance skills by breaking them down into components and utilizing peer feedback for practice and improvement.

- Skill development involves not only instructing, modeling, and observing skills but also nurturing conditional knowledge that enables learners to apply their skills effectively in real-world environments through simulations and scenarios, ultimately leading to successful performance in the workplace.

- Relying on trial and error in addressing performance deficits without a strategic and scientific approach can lead to suboptimal outcomes, such as guessing right by chance, making the situation worse by assuming incorrect deficits, or effecting no change at all.

8

The Deliberate Coach

*If you don't have clarity of ideas, you're just
communicating sheer sound.*

—COMMONLY ATTRIBUTED TO YO-YO MA

There's a wide array of literature on coaching, ranging from evidence-based journals and books by well-known authors to internet articles and posts by experts in the field. If you take a step back and look at the way we're talking about these topics, you'll quickly discover that the discipline of coaching is still developing and lacks a well-grounded definition.

Despite the contradictions, we found that many definitions of coaching overlap. Terms like *supporting, training, partnering, helping, modeling, feedback,* and *goal setting* are commonly used to describe coaching behaviors. And descriptions like *unlocking potential, skill development, goal achievement,* and *performance improvement* are often used to illustrate outcomes associated with coaching.

Many of these same descriptions sometimes overlap with definitions associated with training, managing, leading, and mentoring. In fact, it's not uncommon for coaches to be referred to as trainers or mentors and vice versa. But in the end, it's not about title; it's about impact. We're not concerned with titles so much as the outcomes of the strategies and when they're most valuable. For example, if you apply coaching or mentoring as an intervention when training is needed, the results will fall far short of the desired outcome. Similarly, if you try to engage in managing when leading is needed, disaster can happen. As such, when it comes to improving performance, we think it's important to delineate these concepts so it's clear when coaching is appropriate and when it's not.

Wearing the Right Hat

Leader, manager, mentor, trainer, and *coach* are roles filled by people, whereas *leading, managing, training, mentoring,* and *coaching* are approaches or tools used by professionals in a variety of roles. Like putting on a different hat based on the weather, the most effective leaders understand they must change their approach based on the needs of the people they are supporting (Gavoni, 2024). Each borrows from the same skill sets, and each can be used by a person regardless of their title. And *topographically,* each of these approaches looks similar as it is applied in the moment. For example, each uses feedback as a tool for improvement.

We propose that the difference between leading, managing, mentoring, training, and coaching lies in variables such as timescale and context; however, the most important factor that discriminates between these approaches can be found in the purpose or functionality of each. This is an important distinction because topography and function are very different things. Where topography tells us what a behavior looks like, function tells us why it occurs (Cooper et al., 2020).

Say somebody turns the light switch on a lamp. If the person is turning the light on, their behavior *functions* to increase the amount of light in the room. If they are turning it off, then their behavior functions to decrease the amount of light in the room. In both cases, their topography or form looks exactly the same: turning the light switch. But each behavior serves a vastly different purpose, or function.

Another behavior with different potential purposes is waving. If somebody sees a friend, they might wave to say hello. If they're at a restaurant and would like to request a refill of water, they might wave as a signal for the server to come to the table. In both cases, waving (the topography) looks exactly the same, but they each function for different reasons.

The overlapping descriptions associated with leading, managing, mentoring, training, and coaching are similar. Many of the behaviors connected with these approaches, such as feedback and modeling, look the same. The differences in function, paired with timescales and contexts, are what separate them. Much like wearing a different hat for different conditions, the Deliberate Coach, regardless of title or position on the organizational chart, understands when to engage in which of these approaches based on the need or needs of the individuals they are supporting. Let's take a deeper look.

LEADING

Leading, when understood through a behavior-analytic lens, diverges from traditional leadership theories that focus on inherent traits or isolated behaviors. This type of leading serves as a motivating operation that alters both the value of a goal and the behaviors directed toward achieving that goal. In other words, effective leading isn't just about what's done but the function it serves—specifically, inspiring individuals to collaborate toward a common objective. For example:

EXAMPLE 1: THE TEACHER WHO INSPIRES OTHER TEACHERS

Mrs. Simpson, a seasoned math teacher, takes the initiative to create an informal support group for new teachers at her school. Rather than just providing lesson plans or teaching methods, she shares stories of how she connects the subject matter to real-world experiences, making it more relevant and engaging for students.

Her behavior has a value-altering effect on the new teachers; they begin to see the broader impact of their roles, not just as instructors but as influencers who can make a positive difference in their students' lives. Mrs. Simpson's actions serve as a motivating operation as they encourage other teachers to explore more engaging teaching methods and to contribute to a positive school culture.

EXAMPLE 2: THE SCHOOL LEADER

Principal Patel recognizes the staff's collective stress due to a challenging academic year and increased workload. Rather than issuing top-down directives, he arranges a series of open forums where teachers and staff can share their concerns and collaboratively brainstorm solutions.

His behavior alters the school community's perceptions of what leadership can look like, and his open approach serves to motivate teachers and staff to contribute ideas for improving the educational environment. This unites everyone toward a common goal of better education and workplace satisfaction.

EXAMPLE 3: THE MAINTENANCE PERSON

Mike, the maintenance person, notices that a section of the school grounds has become littered and untidy, detracting from the school's overall environment. He takes it upon himself to clean up the grounds, and he also puts up artwork made from recycled materials to highlight the importance of keeping the area clean. He even starts a mini Green Club, encouraging students to take pride in their surroundings.

His actions serve as a motivating operation that changes the value of the school environment in the eyes of the students and staff. His work becomes a point of school pride and inspires students to maintain cleanliness, thereby working toward the common goal of a better learning environment.

In each of these examples, the individual—regardless of their formal position—serves as a leader by inspiring others to work toward common goals. They alter the value and likelihood of beneficial behaviors occurring in their environment, embodying leading in its truest sense.

Whereas trait-based and other leadership theories may overlook the context and zero in on static characteristics, the behavior-analytic perspective identifies how one's actions in the given environment can motivate beneficial behavior aligned with collective goals. This approach to leading emphasizes the function and impact of behavior within an organization, not just the form it takes or the title held by the individual doing the leading.

The focus here isn't on tasks, metrics, or resources. Instead, it's on the creation of conditions that make beneficial behaviors more likely to occur, naturally aligning those behaviors with broader organizational or group aims. This shift in focus from mere actions or traits to the function of those actions within a specific context makes behavior-analytic leading a powerful way to inspire individuals and drive collective action.

MANAGING

Managing, as understood through a behavior-analytic lens, is not about titles or hierarchical positions, but about the functional role one plays in systematically aligning existing contingencies to guide behavior toward common objectives. Much as the painted lines on the road, the guardrails, and the presence of law enforcement impact driving behavior, managing serves as operational contingencies that channel efforts in the most effective and efficient directions.

In a classroom context, managing could manifest when a teacher adjusts the seating arrangement to minimize distractions, thereby enhancing the students' focus on the lesson. The teacher isn't inspiring students to find value in education (that would be leading); instead, they're rearranging (managing) the existing conditions to facilitate better learning outcomes.

Similarly, in a collaborative project among teachers, managing might be the act of setting up a shared online document where everyone can log their progress, ask questions, and offer resources. This organizational strategy becomes the contingency that shapes and guides collective behavior toward project completion.

School maintenance workers participate in managing by strategically placing wet floor signs and other safety indicators around spills or construction, thereby steering foot traffic in a manner that minimizes accidents and disruptions.

In each of these examples, managing is at work, structuring the environment in ways that naturally guide and maintain behavior toward desired ends. It operates in the background, subtly but significantly impacting the course of action without necessarily altering the value or desirability of the objectives themselves. It's about the effective and efficient journey toward a goal, framed by the strategic arrangement of contingencies.

MENTORING

There are many definitions of mentoring, typically reflective of a personal, long-term relationship that deepens over time and has a ripple effect on reaching personal and career goals. From an academic organizational perspective, mentoring involves a seasoned organization member, often referred to as a mentor, offering guidance, expertise, opportunities, and support to a less experienced, typically younger, individual, known as a mentee or protégé. This relationship is sustained over a period deemed necessary to foster the protégé's development and progress (Erdem & Aytemur, 2008).

Where training is often a formal time-based and skill-oriented process that functions to develop specific skill sets, mentoring is relationship oriented, typically functioning to accomplish general long-term goals related to personal or professional success (Erdem & Aytemur, 2008; Megginson & Clutterbuck, 2005).

Mentoring can be both formal and informal. In either case, it often involves establishing long-term goals, determining actions, reviewing progress, and making adjustments. For example, in education, a seasoned principal might use mentoring to help a teacher who has the goal of becoming an assistant principal improve her leadership skills.

Good mentors employ a variety of strategies for helping their mentees, including training if they see the need and possess the prerequisite knowledge and skills. Similarly, good mentors might serve as coaches and use coaching to help their mentees accomplish goals. We'll talk more about coaching shortly. But first, let's take a functional look at training so it can be differentiated from mentoring and coaching.

TRAINING

Training is commonly understood as a process intended to develop competency of a defined population for a specific purpose. *Cambridge Dictionary* defines it as "The process of learning the skills needed for a particular job or activity" (Cambridge University Press & Assessment, n.d.-b).

In general, training occurs over an allotted time period in a designated area or areas. It focuses on what needs to be done and why it needs to be done, and it involves strategies such as instruction, modeling, and rehearsal to meet a standard of proficiency. From a functional perspective, training is skill-oriented and *functions* as a means of helping a performer develop a skill to the point that it can be demonstrated proficiently and independently.

For example, if a teacher is learning the theory and techniques for asking rigorous questions, she might receive training during a professional development day in the school's designated training area. The teacher's goal is to acquire the knowledge and skills to correctly ask rigorous questions within the classroom. Under these conditions, the trainer might explain the theory behind rigorous questions and academic achievement, model how to ask questions, and require participants to practice by asking questions with a partner as they are provided with feedback. Ideally, this occurs until the teacher is able to perform the task independently.

Once this happens, the teacher is ready to be coached. But what is coaching from a functional perspective, and when is it appropriate?

COACHING

The term "coaching" refers to "the job or activity of providing training for people or helping to prepare them for something" (Cambridge University Press & Assessment, n.d.-a, Definition 5). In the business world, it's been described as "a training method in which a more experienced or skilled individual provides an employee with advice and guidance intended to help develop the individual's skills, performance, and career" (Society for Human Resource Management, n.d.).

If these definitions sound similar to those we shared regarding mentoring and training, it's because they are. This is part of the problem. As you can see, coaching is often conflated with concepts associated with training (e.g., skill acquisition) and mentoring (e.g., career development). While these definitions and descriptions aren't bad, they muddy the water.

While training helps *develop* skills, coaching—as we define it—involves time-based, task-oriented behaviors that *function to transfer a performer's previously acquired skills into the natural environment.* In other words, coaching picks up where training leaves off. When successful, coaching results in sustainable performance and achievement of targeted outcomes. As such, its timescale can be far shorter than that of mentoring.

Coaching can come in different packages depending on the conditions and the needs of the performer. For example, it might include immediate feedback, video feedback, written feedback, modeling, measurement, and so on.

Where training and mentoring tend to occur in sterile environments outside the natural setting, coaching can occur under both simulated (outside the natural setting) or in vivo (live) conditions as part of the transference process. From a behavioral perspective, coaching should focus primarily on successfully applying skills within the work environment and helping performers get in contact with naturally occurring positive reinforcement through feedback and measurement. For example, it might help a teacher to see that students are learning more, students are misbehaving less, or the job has become easier. This will increase the likelihood the teacher will use this skill in the future. However, if a performer does not yet possess the prerequisite skills required to complete the task that would get them in touch with naturally occurring positive reinforcement, then coaching isn't the appropriate intervention—training is.

The Deliberate Coach in Action

In the competitive realm of combat sports, the role of a Deliberate Coach extends beyond training and coaching; it encapsulates leading, managing, and mentoring as well, all tailored to the specific needs of the fighter. Here's a glimpse into this multi-faceted approach:

Inspiration to Compete: Before any training begins, the coach wears the leading hat, igniting the spark that motivates a novice to step into the ring or inspiring a seasoned fighter to aim for a championship title. This initial burst of inspiration is crucial as it sets the stage for all the hard work that will follow.

New Skill Sets: Once the fighter is inspired, it's time for hands-on training. The coach now wears the training hat, establishing the instructing, then modeling, and then requiring the fighter to practice through bag work, mitt drills, and shadowboxing. This structured practice environment is critical for learning new skills safely and efficiently. The coach then switches to the managing hat to ensure the fighter is working through the routines on a schedule.

Emerging Skill Sets: As the fighter exhibits improvement, the coach implements controlled sparring scenarios, getting closer to actual fighting conditions. Here, the coach puts on the coaching hat to ensure the fighter gets in touch with naturally occurring consequences such as getting hit less, landing more strikes, and expending less energy.

Proficiency With Skill Sets: When the fighter reaches proficiency, the coach extends the fighter's skill set by introducing more complex sparring conditions. The focus shifts from immediate technical improvement to long-term mastery and emotional resilience—the cornerstones of a mentoring relationship.

Throughout this journey, the Deliberate Coach adapts their role based on the fighter's needs. This comprehensive, function-based approach ensures not just skill development but also long-term success, providing an all-encompassing support system for the fighter.

In an educational setting, the role of a school leader as a Deliberate Coach is complex, multifaceted, and highly dynamic. The Deliberate Coach moves fluidly between different modes of action: training, leading, managing, coaching, and mentoring, all aimed to meet the evolving needs of teachers. Here's how these roles manifest:

Leading: Initially, the school leader takes on the role of inspiring, aiming to elevate teachers' aspirations. Whether helping a new teacher envision a successful first year or motivating a seasoned educator to become a department head, the act of inspiring sets the stage for what is to come.

Training: The Deliberate Coach then transitions into training, where intensive workshops and professional development sessions take place. This phase equips teachers with critical classroom management techniques, curriculum design skills, and pedagogical approaches.

Coaching: Once teachers show proficiency, coaching begins. At this stage, the Deliberate Coach assists teachers in refining their skills, adapting them to various situations, and making connections to the impact of their behavior. Coaching is especially valuable for ensuring that teachers' skills generalize into lasting, effective practices.

Managing: As teachers become proficient, the act of managing comes into play. The Deliberate Coach ensures that systems are aligned, resources are allocated, and teachers are staying on track toward the established educational objectives.

Mentoring: The Deliberate Coach looks toward the future, focusing on long-term career progression and, perhaps, leadership roles for the teachers themselves.

By acting as a Deliberate Coach, the school leader moves beyond static roles to engage in a dynamic interplay of training, leading, managing, coaching, and mentoring. This function- and outcome-based approach offers a more holistic and responsive model of support, in contrast to traditional leadership theories that rely solely on fixed traits or behaviors. In doing so, the Deliberate Coach paves the way not only for skill acquisition but also for sustained growth and fulfillment for teachers.

Simulations to Foster Coaching Interactions

While the Deliberate Coach understands the need to put on different hats based on the needs of those they are supporting, this book is specifically about supporting the transference of learned skills into the natural environment. In short, coaching. Thankfully, there are antecedent strategies that can be applied to set up successful coaching.

Let's pretend you have the ability to take a fictional trip back in time to a fictional sporting event. As a huge basketball fan (bear with us here), you decide to travel to 1891 to witness the first basketball game as well as the preparation for it. So you put in the appropriate date and coordinates, then find yourself transported to Springfield College, the birthplace of basketball (Naismith, 1996). After doing a little inquiry, you discover that two teams will be playing, so you make arrangements to observe them as they prepare.

The first team, the Memorizers, is busy completing multiple drills to perfect shooting and drilling ability. You watch as a group of young men line up to take shot after shot from a designated spot and are impressed by how often the ball goes into the basket (literally a peach basket). In another group, young men are taking turns running up and down the court and dribbling a large round ball. Their movement is primitive compared to the NBA players of today, but you appreciate their dedication to developing their skills, as you believe practice makes perfect, and you marvel at the high rate of shots the players are making.

After two hours of watching nonstop training on various skills, you venture out in search of the opposing team, the Simulators. As you find a spot to observe off to the side, you're struck by a stark difference. While one group of players is deliberately practicing skills as the Memorizers did, the other group is playing what looks like a practice game on the other end of the court as a coach circles like a shark and gives various players feedback.

You watch them run around awkwardly (by today's standards), missing far more shots than the groups you've witnessed standing at the line making practice shots, but still making a fair amount. As you think about it, you're surprised you didn't observe this method while watching the other team. After all, scrimmaging is a staple of modern-day sports.

The night of the game arrives, and you find yourself a spot on the wooden bleachers. As the teams come out, you wonder who will win the game. While the Simulators were able to scrimmage, the Memorizers engaged in far more repetition and appeared to have greater expertise, as their dribbling and shooting were more precise. This should be interesting.

The whistle blows, and the teams begin making their mark in history. The Memorizers are able to secure the ball first. The player dribbles down the court using the clean and precise technique all of the players displayed during their practice. The Memorizers fans scream in anticipation as he makes his way past half court. And then, suddenly, the Memorizers fans are silenced as the Simulators player steals the ball and passes it to his teammate, who then dribbles it down the court and scores the first point ever. Despite a steady stream of directives from their coaches, the Memorizers are never able to get into the game. Fast-forward to the end, and the analog scoreboard reads 112–4.

So what happened? Well, since you've been reading this book up until this point, we think you've guessed it. While the Memorizers were able to acquire necessary skills, the skills were never properly transferred to actual play. While they effectively developed procedural knowledge (doing the right thing) through lots of repetition, they were never provided the opportunity to develop their conditional knowledge (doing the right thing *at the right time*).

We've been pretty clear about this. Coaching is about getting skills to transfer to the natural environment. Before one can be concerned with transfer, these skills must first be developed, and the Memorizers put a lot of effort into skill development through practice. However, the coaching they received during the game proved fruitless as they'd never engaged in the game-like practice employed by the Simulators. Though well-intentioned, the Memorizers' coaches never provided them with the opportunity to rehearse their skills under simulated conditions that mirrored actual play.

As we've noted, there has been a call to action by many professionals who say we need to develop educators. In our eyes, one of the most effective ways to do this would be housing teacher-preparation programs in actual schools, allowing would-be teachers to practice critical instructional behaviors extensively, and then pairing them with master teachers who are trained in coaching and who can help them transfer their skills into the classroom. Since this isn't likely to happen in the near future, the next best thing might be to create school- and classroom-like scenarios that allow educators to gain repetition in problem-solving, decision-making, and the application of these skills during simulations.

A simulation, or the imitation of the operation of a real-world process or system (Banks et al., 2010), has been used across industries for years to develop effective performance in the workplace. Professionals in economics, medicine, flight, military, sports, and even weather employ both analog and computerized simulations to create conditions that allow for essential practice in a wide range of problem-solving and decision-making processes that they'll face in their respective fields.

Researchers have found that well-designed simulations improve decision-making and critical-thinking skills while they teach discipline-related concepts (Tompson & Dass, 2000). Moreover, participants in simulations frequently report high levels of engagement when simulations employ cooperative learning strategies (Johnson & Johnson, 1998).

Thankfully, simulations are increasingly being used in education, providing the opportunity for teachers and leaders to fill relevant experiential gaps by putting them in virtual school or classroom situations (Badiee & Kaufman, 2014; Cruickshank, 1988). Much can be simulated, including how to respond to student questions, how to break down a lesson, and how to identify the cause for low student achievement. Like scrimmaging in sports, good simulations allow educators to practice while experiencing the consequences of a wide range of decisions. Moreover, they provide the opportunity for coaching to occur, as participants can apply skills they've learned and receive feedback aimed at improving their application in the school.

How do you create a good simulation? It's actually not that difficult. While developing simulations is beyond the scope of this book, it starts by putting together a scenario. Here are a few more tips to get you started:

- Use varied settings.
- Choose meaningful characters and determine role expectations.
- Determine the goals (i.e., what skills you want participants to use).
- Determine how to deliver feedback.
- Establish measurable criteria for success.
- Create branching scenarios to illustrate consequences of decision-making.
- Determine how to debrief.

We hope that we've piqued your interest in simulations. If you wish to explore the topic further, a quick internet search will yield lots of literature.

In Section 2 of the book, we have discussed the science of human behavior, how to build performance, the effective and ineffective use of training, and what it looks like to be a Deliberate Coach. In Section 3, we will dive into the intricacies of Deliberate Coaching, and then provide you with a systematic approach we use in our consulting businesses that will give you a clear path for helping you generalize what you've learned into the school setting.

Key Takeaways

- Seeking to understand, as a Deliberate Coach, the process of identifying the function or purpose behind various approaches to improving performance such as leading, managing, training, mentoring, and coaching is critical. While these approaches may share similar tactics and even appear similar in the moment of their application (the topography of the behavior), their underlying function or purpose (why the behavior occurs) can differ significantly.

- Leading, within the Deliberate Coach framework, goes beyond titles and focuses on inspiring collective action toward shared goals, thereby creating a culture that drives positive outcomes.

- Managing, as defined by the Deliberate Coach, involves strategically using existing systems and structures to guide behavior effectively toward goals, offering a more dynamic approach than traditional situational leadership styles.

- Training in this paradigm aims not just for skill acquisition but for the practical application of those skills, linking them directly to functional outcomes.

- Mentoring is seen as a functionally driven, long-term interaction that focuses on achieving broader personal and professional goals through strategic reinforcement.

- Coaching serves as the centerpiece of the Deliberate Coach approach, aiming to generalize competencies across different settings by aligning them with naturally occurring reinforcement, setting it apart from the competency-building focus of traditional situational leadership.

- Creating effective simulations serves as a crucial tool in the Deliberate Coaching approach, as they provide a structured environment for the effective transfer of learned skills to real-world scenarios.

SECTION 3:

The Deliberate Coaching Model

9

Deliberate Coaching

Sow a thought, reap an action; sow an action,
reap a habit; sow a habit, reap a character;
sow a character, reap a destiny.

—STEPHEN COVEY, *THE SEVEN HABITS OF HIGHLY EFFECTIVE PEOPLE*

Welcome to Section 3. We know we've thrown a lot at you. We also understand that change takes time, and we appreciate the need to start small. Find what will work for you to kick things off, knowing that you will most likely need to demonstrate the impact of this approach on faculty and even demonstrate the coaches' performance to get large-scale buy-in. And that's fine, because this system is based on proven principles that have improved performance at schools, businesses, and countless other places for years. While you shouldn't expect yourself to remember everything you've read, we believe the stage for the Deliberate Coaching approach has been set. In this section, we'll introduce you to a systematic approach and an arrangement of tools rooted in the behavior science that you learned about in Section 2. These tools will provide you with a clear path to accelerating teacher performance at the individual, group, and organizational levels.

Before we dive into the tools, let's take a look at a brief story to review some key concepts. As you read the story, put your behavioral detective hat on. See if you can find critical elements along the way that lead to the success of one group over the others. Moreover, reflect back on what you've learned in Section 2, and try to figure out exactly why these elements help the group be successful. At the end of the story, we'll highlight these elements and give you a brief explanation that should serve to prime you for the rest of the book.

Climbing Mount Achievement

It's the end of June in North Carolina, and the sun slowly begins its ascent on the horizon. A small mountain sits prominently in the background as participants of Camp Achievement huddle around the base, chatting about various topics. The campers are marveling at what they've done so far, happy to have been chosen by their company to participate in this trip as a reward to its top salespeople. They feel fortunate that they were able to bring their family to the all-inclusive resort. Many are reflecting on their experiences in a 4-week training course on hiking and mountain climbing. Most appear a bit anxious as they check their equipment, some seem to project confidence, and one group almost looks indifferent.

Today will mark the culmination of their hard work: reaching the peak of Mount Achievement, a task that has been completed by two-thirds of the previous groups who have attempted the trek. Their instructors have done a fine job training them. Through regular instruction, modeling, and rehearsal with feedback, the group as a whole has expressed confidence that they will reach the top easily. The trek should take approximately 20 hours with travel and rest if the group decides to follow the most frequently used path, though the instructors note that one group a couple of years ago made it to the top inside of 15 hours.

The instructors have divided the group into three teams: Team Chill, Team Quick, and Team Deliberate. Each team has been provided with all the necessary maps, equipment, and supplies. Based on the observations of the instructors, one member from each team has been appointed as the team leader. The team leader's main job up to now has been to facilitate planning meetings in preparation for the trek as well as to act as a guide to ensure a safe journey up the mountain.

During these planning meetings, there were some stark differences between the teams. While the instructors had allotted 4 hours each week to plan, both Team Quick and Team Chill used very little of their planning time. They just knew they were ready. In contrast, Team Deliberate was a little more skeptical. While they believed they could accomplish the task, they felt it would be safer to do a bit of preparation. As such, they could be seen reviewing maps, calculating times and distances, and even placing pins on the map to indicate landmarks. Based on the 20-hour average, they knew precisely how long it should take to reach each one as an accomplishment on the way to their goal.

It's 7:00 a.m. The head instructor blows the horn to signal the start of the journey and tells everyone he'll see them in the main house to celebrate in 3 days. He also reminds them that if they have any trouble, they shouldn't hesitate to use the fully charged radios each of them has been provided with.

It's apparent that each team has planned a different path up the side of the mountain as they all head in different directions. Team Chill's members, looking like seasoned hikers with their backpacks high on their shoulders, move off on a path leading to the right. They look back, give a final wave, and disappear from the view of the instructors. Out of view, the team leader quickly gathers his team around and says, "Okay, who's ready for a beer?" Everybody laughs, and they set off in the opposite direction of Mount Achievement. You see, Team Chill decided long ago that climbing Mount Achievement is of little value. They'd rather be back at the resort with their families enjoying food and drink. Nobody will know, so no harm done. They head off to return to their loved ones.

Team Quick has a different agenda. This team is highly competitive. Not only do they want to climb the mountain, but they want to break the record. As a team, they're determined to reach the peak inside of 14 hours, an hour under the record. According to the maps, the most traveled path isn't a direct line to the top. So why take it? They've decided to take a less traveled path used by more experienced climbers. This path, they've projected, will shave off enough time to break the record.

Just 2 hours into the journey, some members of Team Quick begin to debate their direction as other paths become available. Unable to agree, the team decides to split up into three smaller groups to travel down some of the paths. Since they have radios, they'll be able to contact one another and reunite the team once one of the groups has some indication that they're on the right path. One member, however, becomes so frustrated with the delay caused by the division and lack of direction that he leaves the group to return to the safety of the camp.

After another hour, all groups from Team Quick become discouraged. Though the path up the mountain seemed very clear on the maps and from the base, actually being on the mountain provides a whole new perspective. Unfortunately, Team Quick thought seeing the peak of the mountain through the trees was a good indicator that they were going in the right direction; however, after reaching dead ends on their paths, they are confronted with the error of their ways. Determined to reach the top, the team decides their best and safest option is to return to the base of the mountain and follow the original path recommended by the instructors. They return, inform the instructors of their delay, and set off again. Team Quick reaches the mountaintop in just over 25 hours.

Team Deliberate is taking a different approach, as they've spent a lot of time purposefully planning based on the information provided by the instructors. They've determined what path they'll take, why they'll take this path, who will carry what gear, when they'll stop for snacks, and when they'll camp for the night. Team Deliberate has even determined precisely what indicators will serve as accomplishments to let them know they're moving in the right direction, as they've found landmarks in the maps and literature that will inform them of their progress up the mountain.

So they set off in good spirits at a measured pace. Not more than 2 hours into the journey, some members debate their direction as other paths become available. The team members, once very confident, are beginning to doubt themselves. After all, this is their first journey. The team leader, noticing that the pace has slowed and many members look discouraged, reviews his map and urges them to continue along the same route. "The big boulder should be just up ahead," he encourages them. "Once we come across it, we can be sure we're on the right path and approximately a tenth of the way up the mountain."

Soon enough, the team comes across the big boulder. Together they let out a triumphant cheer, take their first snack as a celebration of reaching their first goal, and set off with renewed vigor and confidence to conquer the climb to the top of Mount Achievement in 17 hours. It's not record-breaking, but it's far better than average, and Team Deliberate is quite satisfied with this outcome.

This story showcases a path to Deliberate Coaching based on the science of behavior (see Figure 9.1). Let's take a look at a few critical elements by comparing and contrasting the teams.

ANTECEDENTS

As discussed earlier, antecedents come before behavior to get it going. They can come in many forms, such as a comprehensive training or a simple reminder. Once outcomes have been identified, the job of a good coach is to identify the behaviors that will lead to attainment of the desired result, then see if those they are coaching possess the required skills. If they don't, the coach must make use of effective training to develop the skills. In the case of all three teams in the example above, the instruc-

Mount Achievement

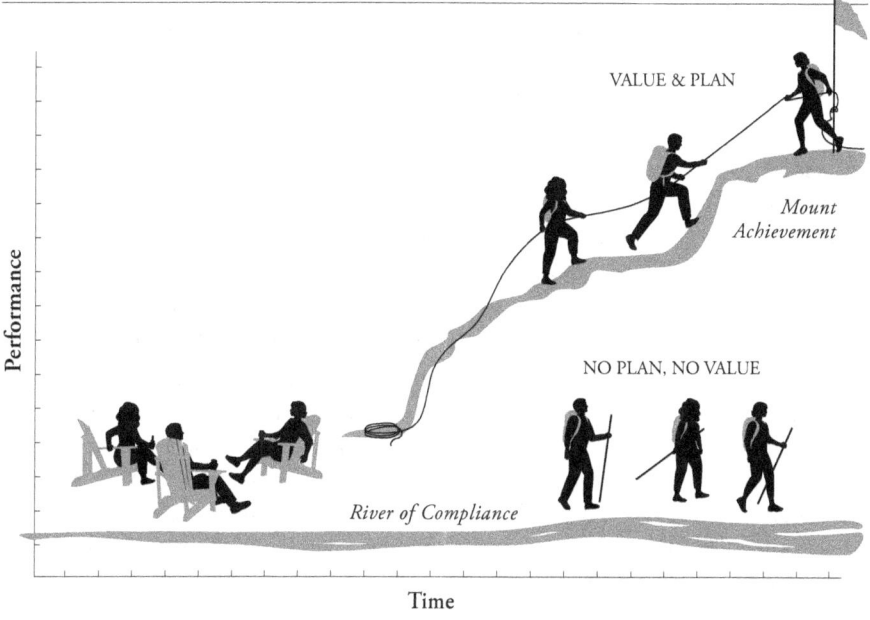

Figure 9.1. *Achieving your goals requires moving past compliance. Teams must work together in a planned and systematic way where everyone is set up for success, encouraged, and can see their own valued impact.*

tors provided instruction, modeling, and lots of rehearsal with feedback to ensure the acquisition of the critical skills. So while not all of the teams reached the top of Mount Achievement, training wasn't the issue.

VALUE

While we know antecedents get behavior going, it is value that keeps it going. Specifically, the value of the consequences that occur as a result of behavior.

In Chapter 5, we talked about positive and negative reinforcement. Reinforcement is at the root of all performance improvement. With the exception of reflexes, all behavior occurs because antecedents start it, and consequences determine whether the behavior will continue. Period.

People do what they do because it works for them. People behave to either get something or get away from something. Consider infants. Why do they cry? While initially a reflexive behavior, infants soon learn that crying sometimes results in access to food or warmth or some other positive experience. Crying is positively reinforced by access to something preferred, and because of this positive reinforcement, the behavior will increase. In other cases, infants learn that crying gets them away from something unpleasant such as a dirty diaper or a loud noise. In this case, the behavior will increase because of negative reinforcement, where crying works to avoid or remove something unwanted.

As adults, we learn that certain series or chains of behaviors can lead to some sort of meaningful outcome. In certain cars, inserting a car key, pressing the brake, and turning the key are steps that will lead to the car starting and your being able to move toward your destination. But the meaningful outcome isn't always provided by the external environment, like the start of the car or feedback from a school leader. The added thing, the meaningful outcome, can simply be feeling good about the results.

A teacher who takes pride in the success of her students is one example. If she attributes some of their success to a certain instructional strategy and this increases her use of the strategy, we can now say that her behavior (using the instructional strategy) has been positively reinforced by naturally occurring positive reinforcement. Nobody had to give her praise, a gift card, or a teacher of the month award. Many teachers teach for the love of helping students. When they see students doing well, this is powerful positive reinforcement.

Now let's get back to our story. If we look at Team Chill, it's clear that climbing Mount Achievement had little value to them. It wasn't a meaningful consequence.

Instead, their behavior was impacted by the use of negative reinforcement—they climbed just far enough to get out of sight, avoiding embarrassment and perhaps the ire of their instructors and fellow group members, then headed back to the comfort of the resort. Remember, this is one of the problems with using negative reinforcement (fear of consequences) to motivate behavior. People will do just enough to get by, and typically only in the presence of the folks who can deliver the undesirable consequence. If people don't value the outcome, they're far less likely to put forth effort toward achieving it.

In the cases of Team Quick and Team Deliberate, it's clear that they both valued climbing to the top of Mount Achievement. Nobody had to look over their shoulder to make them climb. All the behaviors that brought them closer to the top were under contingencies of positive reinforcement. Recall that when people behave in ways that lead them toward what they value, they will continue behaving in those ways, even when nobody is looking.

PRECISE AND PURPOSEFUL PLANNING

While both Team Quick and Team Deliberate valued reaching the top of Mount Achievement, it was precise and purposeful planning that made the difference in their success. Team Deliberate knew precisely how far they would go based on previous hikers' average time and progress. Moreover, they knew precisely what they would see when they got to each predetermined goal. Had they not found landmarks as measurable indicators that they were moving in the right direction, the group might have become disheartened and returned to the base of the mountain, or even to camp, and failed to reach the top. But they didn't leave their journey to chance. They set goals based on the available data, monitored their progress based on the achievement of those goals, and celebrated short-term accomplishments along the way with food, drinks, and snacks.

Your personal journey applying Deliberate Coaching is much like the climb up Mount Achievement. Much like Team Deliberate, you are journeying down a path laden with reinforcement, support, and behavioral principles to accelerate your journey, and the journey of others. There are a lot of moving parts needed to achieve your goals and those of the educators you support. And a successful journey requires collaboration between you and your team of educators and staff. Your school and the interconnected systems must work in harmony to accelerate performance and maximize achievement.

Integration in a Leadership System

There is a subtle simplicity to behavior analysis. Generally speaking, as we've stated before and as a foundation worth repeating, if you see behavior, somewhere in the environment there is at least one antecedent getting that behavior going and at least one consequence responsible for keeping it going. Finding and utilizing these antecedents and consequences, particularly as part of your Deliberate Coaching System, is the cornerstone of effective leadership. Although our assessments and interventions might sometimes prioritize one or the other, it's important that our targeted improvement efforts involve ample amounts of both. As we've hopefully taught you, if all you do is use antecedents such as training, job aids, or prompts without consequences, the best you can hope for is to get behavior started. A behavior won't continue unless something good happens as a result of that behavior. Conversely, if you only focus on consequences, you might be waiting a long time for the behavior to occur. You do need antecedents to nudge behavior in the right direction.

It's important to understand the components of your leadership system in order to ensure that each process is functioning properly and that you have alignment across all of your systems (e.g., human resources, guidance counseling, psychology, administration). The science of behavior is involved before you hire teachers, during the hiring process, during your training and professional development, and through your coaching. Your hiring process involves putting out the expectations for the job and, most likely indirectly through interviews, trying to identify behaviors of applicants that match your expectations. Your assessment of how the repertoire of a given applicant matches the expectations and school culture will help guide your decision-making as a hirer. The training system you use for new hires or professional development will involve a series of antecedents and consequences to build the behaviors targeted by the training. Coaching takes what your teachers learn to another level.

Systems Alignment

Your school leadership system as a whole should contain all of the interconnected processes and systems that all school personnel use to get optimal results. It's important that these systems work together, with each one supporting the others, aligned and interlocked. Each system should also be evaluated as its own entity with its own goals, outcomes, and processes. Ambiguity in your systems can lead to confusion and cause one system to be overlooked or ignored because you might think that another system is meeting those needs.

For example, let's look again at BST, the well-researched training model in the behavioral literature we discussed in Chapter 7. As a system for developing skills, the BST approach consists of providing instruction to a person on the target task being trained, having a trainer model the task, and allowing practice opportunities where the trainee can rehearse the skill and receive feedback.

As a training system, BST is an antecedent designed to help a person acquire skills. But those skills need support in the post-training environment (for instance, the classroom) if they're to continue. A teacher needs to see that what they just learned actually works. This is where your coaching system presumably kicks in. However, a number of coaching interventions in the research literature that are labeled as coaching contain core features of training—namely, instruction, modeling, and feedback (e.g., Komaki & Barnett, 1977; Shapiro & Shapiro, 1985). Do these components mean you actually have a training system rather than a coaching system?

Consider this example:

> A school consultant walks into a school and asks to learn about that school's coaching system, performance management system, teaching evaluation system, or any other system that might help provide day-to-day teacher support. The principal shows the consultant how her staff provides instruction to the teachers when they need help, and how they model the correct way to do things and provide feedback when the teacher practices doing it.

In other words, the principal describes BST. But training is for skill deficits; what system is in place for performance problems that aren't due to skill deficits? You can't fix broken equipment or a procedural issue or a lack of reinforcement with more training. What the consultant wants to know is how the principal helps transfer and maintain skills a teacher learned in college or in a professional-development event. What the principal gave the consultant had more to do with the ways she provides training.

The applied research literature is supposed to provide scientific support and a level of clarification on the form of a strategy (what it looks like) and function (why it's used). While the science of behavior can be a valuable addition to your school leadership infrastructure, behavior coaching has been loosely defined over the years, making it difficult to disseminate proven coaching technologies (Seniuk et al., 2013). If you simply identify a system as "coaching" whenever it contains certain behavioral components, then you might be spending a lot of time trying to fix coaching problems with training solutions. The form might look similar because you use the same laws of behavior to build skills in a training environment as you would to refine and maintain skills in the classroom—but the purpose of training is different from the purpose of coaching. While cars and trains are subject to the same laws of physics, you wouldn't expect a conductor to show you his Ford Mustang® when you ask about how the Amtrak® is running.

Skill deficits should be addressed by a school's professional development and training systems. Transferring those skills and supporting day-to-day teacher performance should be addressed by your coaching system. Systems fixes should be fixed by your administrative systems. While these systems all work together, they serve different purposes. School leaders should know how each system is working and where there are opportunities to improve. If you confuse one system for another, you could be ignoring a critical piece of your school's infrastructure. Training on the effective use of a SMART Board won't do anything if the SMART Board is broken.

Remember that there are many hats a leader wears, and we match our duties based on the needs of those around us. In alignment with the different roles leaders play and the vast expansion of obligations leaders must pay attention to, there are three main categories of leadership systems related to coaching and performance-improvement efforts, as depicted in Figure 9.2.

ADMINISTRATING

A good coach takes time to observe what's going on, monitors progress, and finds ways to set a coachee up for success. Sometimes

Figure 9.2. Leadership is not isolated to simply administrative duties, training, and coaching. It is an ongoing cycle of setting people up for success and providing ongoing support.

the result of an assessment or observation is that a new policy needs to be written for a school or classroom. Maybe you need to order new equipment, build a new checklist to help teachers organize a new activity, or tweak a couple of items in the academic calendar of events. All of these are things you've identified as barriers hindering teacher performance. They're all important duties of a good leader, even if they aren't part of your coaching.

TRAINING

Professional development, continued education, modeling, and other forms of training are critical to ensuring that teachers know what to do. But let's be clear that it's not the *knowing* that dictates the success of training—it's whether a teacher can demonstrate the trained skills at a high enough level that you can reasonably expect them to use them in the classroom.

Regardless of how good your training is (how much *demonstrable* learning it produces), it's still just an antecedent. It will still only get behavior going. You need to provide opportunities to practice these skills in the classroom. And once teachers try them out, good things have to result, or they won't use them anymore.

COACHING

Training and administration have stopping points. Remember, the demonstration of learning is the main goal of training (Johnson & Rubin, 2011; Molenda & Russell, 2006). Someone has a skill deficit, needs training, gets trained, training ends, and the person is presumably good to go. A process needs to be fixed, someone fixes it, the new process gets rolled out, and the new process is in effect. Coaching, however, doesn't stop. Coaching, in particular the type of *behavioral* coaching described here, involves the ongoing use of behavioral strategies to directly impact performance (Weatherly, 2019). Coaching helps facilitate the transfer of skills trained into the workplace environment.

In Chapter 13, we're set to guide you in designing your very own Deliberate Coaching System. However, before we dive into that, it's important to unravel the concept of Deliberate Coaching itself. This goes beyond the conventional coaching methodologies you might already know. Let's explore what makes Deliberate Coaching distinct and why it is imperative to understand its nuances.

Deliberate Coaching Defined

If the school were a movie set, teachers would deserve the leading role. Because good instruction has the largest impact on student achievement, every other adult in the building should be thought of as a supporting cast member, putting teachers in a position where they can shine like the stars the students need them to be. In much of the recent educational literature, there is a call for school leaders to help teachers excel by not only focusing on the managerial demands of school leadership, but also focusing on the specific tasks required for improving classroom instruction (Steiner & Kowal, 2007).

Some school leaders attempt to spread out their leadership responsibilities by delegating leadership duties across personnel. Some leaders seek to embed a coaching program into their school by employing classroom-level instructional coaches who provide teachers with needed support (Spillane et al., 2001). However, it's not as simple as just hiring instructional coaches and hoping for the best.

For a coaching program to be effective, school leaders must develop a coaching strategy as part of their school improvement plan, select the right coaches, evaluate the impact of coaching, and provide regular feedback to encourage and guide coaching efforts. Essentially, the school leader must act as a coach to the coaches.

"Good leaders are like good coaches. They know how to bring out the best in people on their team."

—John Maxwell, *Good Leaders Ask Great Questions*

As we've discussed, deliberate practice refers to practice that is purposeful and systematic. While regular practice might include mindless repetitions, deliberate practice requires focused attention and is conducted with the specific goal of improving performance.

Borrowing from this familiar concept, we describe Deliberate Coaching as coaching that goes beyond the functional definition we've previously outlined.

Deliberate Coaching is a model of behavioral coaching that revolves around precise, purposeful, and systematic coaching interactions. Deliberate Coaching is implemented by internal personnel to directly impact workplace behaviors that are aligned with workplace results. As an approach that's based in research on the science of behavior, Deliberate Coaching requires focused observation; tracking of performance change; and feedback loops to shape socially valid, high-impact behavior. And when we say socially valid, we mean behaviors that are acceptable to both the people implementing a particular coaching intervention (e.g., the teacher) and those receiving the intervention (e.g., the students). Let's take a look at the principles used to guide Deliberate Coaching:

PRECISE

- measurement of performance that is specific, detailed, and frequent
- clear distinction between the development of skills (training) and transference of skills
- feedback that is specific to a pinpointed behavior
- feedback that links behavior to pinpointed results

PURPOSEFUL

- linked to on-the-job outcome
- intentionally designed with positive procedures
- sustained focus on a few pinpointed behaviors
- intentionally designed to be brief

SYSTEMATIC

- uses specific behavioral procedures that have been experimentally demonstrated to be effective and are repeatable
- measures improvement against a performer's own performance
- uses measurement of performance to guide coaching behavior
- uses social validity to monitor and guide coaching interventions

This is not an abstract ideal but a concrete set of practices that can be dissected and understood. To drive this point home, let's provide an example and non-example of Deliberate Coaching in action.

EXAMPLE

Over the course of a week, the school leader conducts frequent, unannounced 15-minute observation periods, focusing specifically on the teacher's questioning strategies. Data are collected on the types of questions posed, the wait time allowed for student responses, and the complexity of student answers. During the weekly coaching session, these real-time data are discussed with the teacher. Using this concrete evidence, the leader, using good questioning strategies, collaborates with the teacher during coaching briefs to pinpoint areas of strength and those in need of improvement. The teacher is then given tailored suggestions for phrasing questions that encourage higher-order thinking, along with strategies for effective wait time and student engagement. Both agree to track a specific set of metrics for improvement, which will be reviewed in the next session.

This frequent and direct observational approach not only supports the teacher's immediate needs but also serves as an ongoing feedback loop for continuous professional growth.

NON-EXAMPLE

The school leader schedules a monthly meeting with the teacher. During the meeting, general topics such as "improving student engagement" or "increasing classroom participation" are discussed, without any specific data or observations to back up the conversation. The school leader offers generic advice such as "You should try to be more engaging" or "Maybe switch up your teaching techniques," without giving detailed, actionable steps. There is no follow-up mechanism to track if any improvements have been made, nor is there any targeted plan for professional development based on the teacher's individual performance metrics.

In this non-example, the coaching is neither frequent nor targeted, and it doesn't engage the teacher in a collaborative, data-driven process aimed at specific performance outcomes. Therefore, it's unlikely to lead to meaningful improvement.

In contrasting these scenarios, the gulf between Deliberate Coaching and its less-structured counterpart becomes apparent. In the example, the school leader uses precise measurements, driven by data and direct observations, to focus on specific, actionable behaviors. This approach isn't just specific; it's systematic and purposeful, tailored to induce not just any change, but meaningful change that directly impacts the teacher's performance and, ultimately, student outcomes.

In stark contrast, the non-example depicts a haphazard approach, lacking both the frequency and specificity of observation. The advice dispensed is generic, and there's no systemic method to track improvements or setbacks, rendering the coaching sessions largely ineffectual. While the latter may be a form of coaching in name, it lacks the focused intent and scientific underpinning that make Deliberate Coaching so potent for fostering real, sustainable change.

Thus, when we talk about Deliberate Coaching, it's not a matter of merely going through the motions. The devil, as they say, is in the details. And it's those details—those pinpointed, deliberate efforts—that spell the difference between mere activity and meaningful accomplishment.

The Deliberate Coaching process is designed to work within the constraints of any environment. Each school has unique parameters that need to be acknowledged and respected. Precision allows for objectivity, taking away subjective judgement that can cloud progress and moving toward focused directives. Purposeful coaching ensures there is a reason behind the coaching that both the coach and the coachee can see. Being systematic ensures you're taking things in manageable increments, monitoring your progress, and making certain efforts are supported and goals are reasonable. These foundational targets can set up any coach for success!

Deliberate Coaching at Scale: Zooming In and Zooming Out

The dialogue surrounding educational leadership has long emphasized the importance of instructional leadership—a model predicated on the belief that direct, individualized coaching on instructional practices is the linchpin of student achievement. This approach, championed by decades of educational scholarship, argues for a hands-on engagement by school leaders in the pedagogical development of their teachers. The rationale is clear: Outstanding instruction is the direct path to elevating student performance.

Yet this model, while compelling in theory, often encounters significant obstacles in practice. The assumption that school leaders can consistently engage in direct coaching overlooks the complex realities and diverse challenges inherent in school management, particularly in underprivileged settings. The expectation for leaders to dive deep into the intricacies of classroom instruction, akin to expecting a general to micromanage the movements of every soldier, often proves impractical amid the broader strategic challenges they face.

Expanding the Scope: Deliberate Coaching at Scale

Realizing meaningful improvement in educational outcomes requires a strategic approach that spans beyond the confines of the classroom to embrace systemic and procedural enhancements. Deliberate Coaching must seamlessly navigate between micro- and macro-level interventions, applying a targeted, systematic methodology that not only elevates individual teacher performance but also addresses the systemic scaffolding that underpins the educational ecosystem.

DELIBERATE COACHING AT THE ORGANIZATIONAL LEVEL

Deliberate Coaching at the organizational level is about the big picture: How do the school's systems and policies set teachers and students up for success or failure? It's less about the moment-to-moment interactions and more about the overall environment in which those interactions occur. Imagine a school that's grappling with chronic tardiness. A leader focused on organizational coaching would look at schoolwide policies, perhaps adjusting the start times, revamping the morning routine, or implementing a reward system for punctuality.

Imagine a school where communication breakdowns between departments lead to duplicated efforts and missed opportunities for cross-curricular projects. In this scenario, Deliberate Coaching at the organizational level would look like a strategic intervention to overhaul and improve the schoolwide communication infrastructure. The coach would start by conducting a thorough assessment of current communication practices and policies to pinpoint exactly where and why breakdowns are occurring. This would likely involve interviews with staff from various departments, surveys to capture everyone's perspective, and direct observation of departmental interactions to understand the existing communication flow.

Once the problems are identified, the coach would work with the leadership team to design a comprehensive communication strategy. This could include establishing clear protocols for inter-departmental collaboration; introducing regular interdisciplinary meetings; or implementing a shared digital platform where teachers can plan, share, and collaborate on cross-curricular projects seamlessly.

The coach would also ensure that there are systems in place for ongoing monitoring and feedback about the new communication processes. This might involve regular check-ins with departments to address any teething issues or resistance to the new strategies, and then adapting the approach based on this feedback.

Additionally, Deliberate Coaching would include training sessions for staff to ensure they are competent in using any new communication tools and that they understand the expectations for collaboration. Throughout the process, the coach would emphasize the value of open, transparent communication and how it can enhance the educational experience for both teachers and students.

In essence, Deliberate Coaching at the organizational level in this scenario is about creating a culture of collaboration through structured, supported, and systematic changes to communication practices across the school.

DELIBERATE COACHING AT THE PROCESS LEVEL

Then there's Deliberate Coaching at the process level. Here we look at the specific workflows and protocols that make up the school day. What procedures are in place, and how do they help or hinder the teacher's performance? It's about fine-tuning these processes so that they better support teaching and learning. At the process level, we scrutinize the mechanisms within which individuals operate. Take Ms. Alvarez, a school administrator who is trying to streamline the process of reporting student assessments. Through Deliberate Coaching, we'd help her analyze the current workflow, identify bottlenecks, and design a more efficient system. This could involve introducing a new software tool or retraining staff on data entry protocols. Or the Deliberate Coach might seek to build up and retain quality teachers by improving induction processes that promote career learning, provide multiple personnel with access to support, and treat induction as part of a lifelong professional development design (Breaux & Wong, 2003).

Continuing with the organizational level scenario, at the process level, Deliberate Coaching would focus on refining and improving the specific workflows and methods that contribute to cross-departmental communication. The coach would closely examine the existing processes—how information is shared, how joint decisions are made, and how cross-curricular planning is documented. For example, they might find that there's no standardized process for sharing upcoming curriculum topics, leading to missed opportunities for creating cross-curricular links.

To address this, the coach would work with the leadership team and staff to develop a clear, step-by-step process that ensures information about departmental plans is shared in a timely and efficient manner. This might involve setting up a shared calendar for all departments to log important dates and curriculum topics, or establishing a weekly bulletin where each department outlines its key learning objectives for the coming weeks. The coach would also facilitate the creation of templates for lesson planning that encourage teachers to think about and document potential cross-curricular connections. They would then train the staff on how to use these templates effectively.

Throughout the process, the coach would encourage staff to give feedback on these new processes, maintaining a flexible approach to fine-tune the workflows as needed. This might include simplifying documentation, adjusting the frequency of shared planning meetings, or streamlining communication channels based on what works best for the teachers. By focusing on these specific, actionable changes at the process level, Deliberate Coaching aims to build a more efficient and effective framework for inter-departmental collaboration that supports the broader organizational goal of a cohesive educational experience.

DELIBERATE COACHING AT THE PERFORMER LEVEL

At the performer level, we've got a good grasp on coaching individual teachers. But let's continue the scenario from above to see how it connects the dots. Shifting the focus to the performer level, Deliberate Coaching becomes even more individualized, homing in on the specific behaviors and skills of the educators within the framework of the new communication strategy and processes.

For instance, a coach might observe that some teachers excel at identifying opportunities for cross-curricular activities but hesitate to reach out to colleagues due to a lack of confidence or uncertainty about protocol. Here, Deliberate Coaching would involve one-on-one sessions with these teachers to reinforce the importance of their contributions and to practice effective communication skills.

The coach would guide teachers on how to initiate and navigate collaborative discussions with their peers. This could involve role-playing exercises to build confidence, providing scripts or language structures to help start conversations, or teaching active listening skills to ensure they can effectively collaborate with colleagues from different departments.

Furthermore, the coach and the teachers would agree on clear, measurable goals such as initiating a certain number of collaborative projects per semester or attending cross-departmental planning sessions. The coach would ask good coaching questions and provide regular feedback, highlighting successes and areas for growth, ensuring that each teacher understands how their individual actions contribute to the wider organizational objectives.

For teachers already skilled at interdepartmental communication, the coach might challenge them to take on leadership roles within collaborative projects, fostering a sense of ownership and responsibility that can spread to influence their colleagues. At the performer level, the key is to empower each teacher with the skills and confidence to act on the new communication processes, ensuring that they not only understand the organizational vision but also feel equipped and motivated to contribute to its realization.

It's not enough to simply give advice and feedback to teachers. Leaders as Deliberate Coaches must develop and implement systems that support effective teaching and behavior management schoolwide. Good coaching across these levels doesn't look the same for every school. In schools where the basics are in place and things are running smoothly, focusing on individual coaching might be enough. But in struggling schools, leaders must also build the systems that support teaching and learning.

In short, Deliberate Coaches need to look beyond individual classroom performance to the broader systems that support it. That's the kind of leadership that moves beyond managing to create sustainable change. It's about being precise, purposeful, and systematic, ensuring that every action taken aligns with the goal of improving student outcomes across the board. As a Deliberate Coach, mastering the art of zooming in and out is vital, as it allows for nuanced interventions at the performer level, targeted refinements at the process level, and strategic overhauls at the organizational level. By adeptly navigating these layers, a coach ensures that individual actions, system processes, and organizational structures are cohesively aligned, driving the school's collective march toward excellence.

Key Takeaways

- Behavior change is most effective when rooted in a blend of training, administration, and coaching—all grounded in the principles of behavior analysis.

- Administrative tasks are not ancillary to leadership; they can act as critical antecedents that set the stage for effective teaching, removing barriers that hinder performance.

- While training is crucial for skill development, it's just the starting point; the ultimate goal is the application of these skills in real-world settings including classrooms.

- Coaching, in contrast to training and administration, is an ongoing process aimed at continuously improving performance by using behavior-analytic strategies.

- Deliberate Coaching is a distinct version of coaching in that it's precise, purposeful, and systematic, with its roots in behavioral science research.

- Specific, frequent, and detailed performance measurement is crucial for effective coaching, providing a robust foundation for behavior change.

- Coaching should be purposefully designed with positive procedures, focusing on a few pinpointed behaviors for sustained impact.

- A systematic approach to coaching utilizes specific behavioral procedures that have been experimentally proven effective, ensuring repeatability and scalability.

- The best leaders serve as coaches to the coaches, embedding a strong coaching program into the organizational culture as part of a broader school improvement plan.

- Social validity, the acceptability of the behavior intervention to both implementers and recipients, is a key metric to consider in the coaching process.

- Deliberate Coaching requires a holistic approach, addressing not just individual teacher performance but also systemic and procedural aspects of education.

- At the organizational level, coaching involves strategic overhauls of systems and policies to create a conducive environment for both teaching and learning.

- Process-level coaching focuses on streamlining day-to-day workflows and protocols to support and enhance teacher performance and cross-departmental collaboration.

- Coaching at the performer level is highly individualized, concentrating on the specific behaviors and skills of teachers within the school's broader operational framework.

- Deliberate Coaches must be adept at zooming in—for detailed, personalized coaching—and zooming out—to align their coaching with systemic and organizational goals.

10

The 8 Ws of Deliberate Coaching

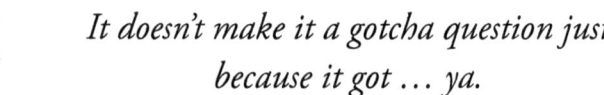

It doesn't make it a gotcha question just
because it got ... ya.

—Jon Stewart, *The Daily Show*, June 6, 2011

Famed leadership guru John Maxwell recognized questioning as foundational to successful leadership. In his book *Good Leaders Ask Great Questions* (2014), Maxwell notes, "If you want to be successful and reach your leadership potential, you need to embrace asking questions as a lifestyle" (p. 4). And we can say the same about Deliberate Coaching. While high levels of practice and feedback typically occur in the early stages of effective skill development, Deliberate Coaches understand that to support the transference of those skills into the natural environment, they should progressively fade from *telling* to *questioning*. And this applies not only to a leader's interactions with those being coached but also to how the leader plans, delivers, and tracks their coaching interactions.

The 8 Ws of Deliberate Coaching

"Plan your work, work your plan." "Measure twice, cut once." "Look before you leap." Whatever cliché you heard the most as a kid, it still applies today. "There are only so many hours in a day," and school leaders "have a full plate." Thus, we're looking for efficient, effective, and lasting coaching. If it's hard to get a moment with a teacher given your schedules, don't you want to make sure you make the absolute most of it?

What questions do you ask yourself when you're coordinating your time and looking to maximize the time you have with your educators and staff?

It's the evidence gathered from what's actually happening that powers up coaching, turning it from guesswork into a focused tool for improvement.

The five Ws (who, what, where, when, and why) are often referenced as a problem-solving guide, originating in concept with ancient philosophers such as Aristotle and evolving over time in applications of science and information gathering (Sloan, 2010). The Deliberate Coaching process is a science-based process that shares a similar need for information gathering and problem-solving. The process must be tracked and built on to ensure that coaches are making a significant impact. The 8 Ws of Deliberate Coaching capture this process and function both as a guide for planning and as a way to report how your coaching interactions went (see Figure 10.1).

8 Ws of Deliberate Coaching	
Who? Who are you coaching?	
Why? Why is this result important?	
Which? Which behaviors are you targeting?	
When? When will you coach?	
Where? Where are you coaching?	
What? What will/did you do?	
What Happened? What was the impact?	
Why Don't … ? What will you do next?	

Figure 10.1. The 8 Ws of Deliberate Coaching offer a way to not only plan your coaching, so as to ensure you're efficiently supporting progress, but also a way to track what happened and what you'll do next once you put your plan into action.

WHO?

Who are you coaching? Although there is a time and place for group discussions, coaching interactions are a one-on-one exchange. What's reinforcing for one person might not be reinforcing for another person, and what's causing a performance problem for one person might not be an issue for someone else. Because of this, you should make plans for and report on one person at a time.

EXAMPLE

Put yourself in the position of a vice principal responsible for a team of teachers. You will need to identify one teacher per 8 Ws form. The goal of this form is to plan, track, review, and systematically move forward with an individualized approach to teacher support.

WHY?

Why is this important? The answer to the "why" is the result or outcome you're looking to impact with your coaching. This is how you know there is something that requires coaching support and how you'll know that your coaching is working. Results should be objective, observable, and easily measurable so you, the coachee, and all applicable parties can see and support progress.

EXAMPLE

Ms. Reid is a new teacher, and her students have been scoring low on their math tests recently. The test scores in Ms. Reid's classroom are lower than standards and expectations for her grade. Math is an important area in subsequent grades and in school standardized assessments, so it's imperative that these scores come up. Ms. Reid has the same tools and materials as other teachers, but we need to get to the bottom of what support is needed and the behaviors that need to be coached. You need to justify to yourself, your teacher, and your supervisors why you selected to coach the particular behaviors you're coaching. Your time is valuable, and what you do with that time is important.

WHICH?

Out of all the behaviors that lead to those results, which behavior(s) are you targeting with your coaching? There must be alignment between behavior and results, and coaches must monitor both to get change and get it the right way. And these pinpoints need to be clear to all applicable parties. The objectivity of your pinpoints is essential to ensuring that you know what you're focusing on, you know what to track and how to discuss progress, and you can provide a level of clarity during feedback discussions. This is the information that needs to be provided to your coachees while you're tracking where they've been and where they're going. Remaining transparent and clear along this journey will help you gain credibility as a coach. You want those you are coaching to trust you. You want them to *want* to be coached.

EXAMPLE

Given Ms. Reid's low test scores, you've been spending time to better understand this problem. While observing and discussing this with Ms. Reid and exemplar teachers, it becomes apparent that she runs through her fluency flashcards slowly, losing her students' attention in the process. Getting her to increase attention by using the flashcards quickly should help improve her students' scores.

WHEN?

When is the best time for your coaching? When does the behavior occur most frequently? When are there opportunities for the behavior to occur? Most likely your school won't have a staff of free-roaming coaches, meaning that only a limited number of school leaders will be functioning as coaches. Your school will need to allocate coaching wisely if each teacher is to get the support they need. One way to do this is carefully planning when you're going to observe, when you're going to assess, and when you're going to have your quick coaching interactions. Optimally, you'll coach where the work takes place (e.g., in the classroom), respecting privacy and confidentiality and ensuring close proximity to when the behavior is occurring; however, this isn't always possible. Use remote coaching in these situations, focusing on open-ended coaching questions and quick concise interactions, helping the individual talk about their behavior and impact in a way that paints a clear enough picture for you to coach.

EXAMPLE

Ms. Reid's math activities occur at 10:30 a.m. and 2:00 p.m. The best time to offer support would be during or immediately after these activities. You don't want to interrupt the lessons, but if you observe and then provide feedback directly once school ends, this should allow you to gather the information you need for your feedback and to give Ms. Reid the privacy she needs when you discuss her performance.

WHERE?

Where will the coaching take place? This is directly linked to the "when" component in that you're trying to maximize the time you're spending on coaching in an agreed-upon area that is conducive to coaching interactions.

EXAMPLE

Because the coaching interaction will come after the students have left, it will take place in Ms. Reid's classroom.

WHAT?

Based on what you know about a particular educator's performance, what do you need to do as a coach to move forward? What are you *planning* to do? Afterward, what did you *actually* do? (Hopefully it's the same as what you planned, but sometimes things come up.) Here are some potential things a coach might do, all of which are part of or at least linked to the coaching interaction.

Observation

In order to coach, you need to know what's happening with the person you're coaching. You can't coach without pinpointing behaviors and results, and you can't coach without understanding what's going on around the behavior. To discover these things, you need to look around and observe what occurs before, during, and after the behavior. If you're trying to help a teacher improve student participation, for instance, it's good to see what the teacher is currently doing to encourage that participation and what conditions surround these behaviors. When planning your coaching, sometimes the coaching behavior you're looking for on a given day is focused on assessment by observing to find the "which" and the "what." For example, let's take a quick look at the power of observation.

> In Mrs. Ellis's fourth-grade class, student participation has hit a plateau, despite her earnest efforts. The enthusiasm once evident is dimming. Enter Mr. Park, the instructional coach, with a clear objective: boost student engagement.
>
> On a typical Wednesday morning, Mr. Park positions himself discreetly at the back of the class. His goal isn't to evaluate Mrs. Ellis but to understand the "what" and "which": what she is doing to encourage participation and which conditions are influencing the current classroom dynamics.

As he observes, he takes note not only of Mrs. Ellis's strategies but also of what happens right before a student decides to engage or withdraw. Is Mrs. Ellis acknowledging every hand raised? What is her response when a student gives an incorrect answer? How do the students react to these outcomes?

Mr. Park notices patterns. Participation spikes when Mrs. Ellis uses visuals and wanes during long stretches of lecture. Students seem eager to speak when discussion follows a group activity but are reticent when called on individually.

Armed with these observations, Mr. Park plans his coaching session. He isn't there to overhaul Mrs. Ellis's approach but to fine-tune it, to shift her focus toward what is working and away from what isn't. By understanding the environment and the behaviors within it, Mr. Park is ready to help Mrs. Ellis adjust her methods so that student participation will once again flourish.

Mr. Park's approach highlights a simple truth: Careful observation leads to better coaching. By looking for patterns in the classroom, he is able to give Mrs. Ellis clear, actionable advice. It's the evidence gathered from what's actually happening that powers up coaching, turning it from guesswork into a focused tool for improvement.

Assessment

Deliberate Coaching is purposeful coaching. Your coaching interactions must serve a purpose, and this purpose should not be a guess. An efficient coach maximizes the coaching interaction by using behavior science to assess why bad things are happening and why good things are happening. (Don't make the mistake of ignoring the good things, or they might go away!) Using tools such as the Performance Diagnostic Checklist (see Chapter 12 and Figure 12.1) to assess why behavior problems are occurring, and identifying antecedents and consequences surrounding desired and undesired teacher performance, will help you find the right coaching solution for the situation. You don't want to recommend training if it's not a training problem, and you don't want to waste time with constructive feedback if the problem is a bad process or faulty equipment.

Trial and error is costly and time-consuming, so sometimes the best thing for a coach to do is assess. Then you can use the information you gather to guide the coaching interaction. Let's look at the power of performance diagnostics in action!

In the bustling environment of Jefferson High School, Coach Taylor has a clear objective: to assist Mr. Mueller, a seasoned history teacher, whose class recently seems to be drifting into disengagement and lackluster performance. The dip in student interaction is notable, and both student grades and Mr. Mueller's morale are beginning to show it.

Instead of jumping to conclusions or rolling out generic advice, Coach Taylor decides to take a step back and assess the situation meticulously. She spends days in Mr. Mueller's class, not as an overseer but as an observer, using the Performance Diagnostic Checklist as her guide. She is determined to pinpoint the precise factors contributing to the lackluster classroom dynamics.

She notes how Mr. Mueller's usually vibrant discussions have turned into monologues. She observes the timing of his lectures, the density of the material, and the students' responses. She pays close attention to what happens right before the energy in the room wanes—is it a certain teaching method, or perhaps the content itself?

The checklist helps her realize that it isn't a lack of skill on Mr. Mueller's part; the issue is that his tried-and-true methods aren't resonating with this particular set of students. The problem isn't one that requires more training for Mr. Mueller; instead, it needs a shift in his approach to match the changing needs and interests of his students.

Coach Taylor uses her findings to craft a coaching plan that is as precise as it is practical. Together, they introduce more interactive elements into Mr. Mueller's lessons and align his strengths with the students' needs. The impact is evident within weeks. Student participation bounces back, and the grades follow suit.

This narrative exemplifies the essence of Deliberate Coaching—assess first, then act. Coach Taylor resisted the urge to rely on assumptions or past experiences alone, allowing for a thoughtful assessment that led to a targeted and successful intervention. It is a win for the students, a win for Mr. Mueller, and a testament to the power of a well-informed coaching approach.

Prompting

Behavior needs something to get it going. Optimally, these antecedents occur naturally in the environment so that a teacher is simply reacting to normal classroom events.

- The time of day prompts the teacher that a new activity should begin.
- A student's question prompts the teacher to address that student's need.
- If a particular student has been struggling with multiplication, then a new multiplication activity should prompt the teacher to check on that student.

But sometimes people need a nudge. Your coaching interaction could offer some quick and well-timed advice: asking the teacher to discuss the safety procedures prior to leaving on a field trip or reminding the teacher about what seemed to work the last time during a particular activity. You know that the teacher *can* do the skill, but sometimes a prompt is the type of antecedent they need to get things going. Let's peek in to check out Coach Williams effectively using prompting with a teacher.

> In the organized chaos of Mr. Garba's classroom, with a field trip on the horizon, excitement is the order of the day, threatening to eclipse the critical safety briefings. Coach Williams, understanding the dynamics of this environment and the demands on a teacher's attention, sees an opportunity for a strategic intervention.
>
> She approaches Mr. Garba with the kind of straightforward, actionable advice that comes from a place of experience. "Think about running through the safety procedures before the buses roll up. It worked like a charm last week," she reminds him.

It is a small, yet pivotal, prompt—a necessary antecedent to trigger the desired behavior in Mr. Garba. It isn't micromanaging; it is a timely reminder, a coach's nudge to ensure that important protocols aren't lost in the shuffle. Mr. Garba appreciates the nudge, realizing that in the whirlwind of his responsibilities, these prompts are not an intrusion but an asset.

Feedback

Feedback is information about what the person is doing and is targeted at improving performance. It can be positive, highlighting what someone is doing well and why it's important, or it can be constructive, identifying what the person is doing wrong, what should be done instead, and why this change is important.

Each type of feedback can be used by a coach both as an antecedent to get behavior going or as a consequence, depending on how and when it's delivered. Suppose a teacher is typically struggling to maintain order when her class is walking to lunch. If you provide the teacher with feedback *before* lunch, the feedback is an antecedent; your intent is to guide her upcoming transition. If you provide the teacher with feedback *after* lunch, the feedback becomes a consequence, helping to maintain the teacher performance. In this case, the feedback could actually be working as a *dual-functioning chained stimulus* in that it serves two purposes: to reinforce the response that came before it (what the teacher did correctly during the transition to lunch) and prompt the response that comes next (what the teacher does during the transition from lunch to the classroom).

This is a good way to use constructive feedback, as it's critical to provide positive feedback once the teacher engages in the new preferred behavior. This shows the teacher that you're paying attention when good things happen too, and you don't just pop in for the bad things. It helps build trust, showing that if you say you'll follow up, you actually do follow up. And it gives you the opportunity to show that there is value in doing things the preferred way, both in terms of how it impacts the teacher's work and in terms of the positive reinforcement you offer as a coach.

In order to provide feedback in the kind of specific and individualized way that makes it effective, you need to know what results and behaviors you're targeting, why the performance is where it is, when the best time is to offer feedback, and the other items in the 8 Ws. You might be able to find some of this information by asking good questions during the coaching interaction. If not, that's where the observation and assessment come in. Although performance feedback is a critical piece of the coaching interaction, don't neglect the observation and assessment that help maximize the impact of your feedback.

Maximize the efficiency and impact of your coaching, and PIVOT how you're planning and delivering feedback. The PIVOT Feedback approach offers a useful toolbox to guide feedback interactions. The approach, consistent with the standards of behavior analysis offered in the Deliberate Coaching process, maintains a focus on individualization and consistency. PIVOT Feedback is:

- **Precise**
  - There is a level of trust between a coach and the person being coached. This trust is built in part through establishing clear pinpoints and providing accurate feedback related to them. Without precision and accuracy in feedback, the coachee and the coach won't be on the same page. Imagine you're a teacher, and your principal comes in to give you feedback on the low scores in your classroom, but the data she's reporting to you are wrong—they're from a different classroom. This error won't exactly build confidence in the feedback to follow! Precision helps build trust because it feels good to know with reliability that you and the coach are talking about the same thing.

- **Individualized**
  - Results come through people, and each individual person deserves support. Feedback should be individualized to the person, genuine, and sincere based on what this individual is doing in their individual environment, focusing on what's working, what's not, and why this is important to the individual.

- **Valued**
  - People do what they do because it's working for them. For feedback to be effective, they must see the value in adhering to the feedback. Good coaches help make this link between behavior and impact. And remember,

always put yourself in the performer's shoes—is it making the link to what is of value to *them*?

- **Ongoing**
  - Feedback is ongoing: That is how behavior is shaped, by acknowledging incremental improvements toward a goal and following up when an undesired behavior starts to go away and a better behavior takes its place. Feedback needs to occur frequently enough to shape behavior and support those subtle changes. There is nothing worse than receiving constructive feedback on a behavior that has worked for you, going against the grain and trying the replacement behavior discussed during the constructive feedback conversation, and being left hanging with no acknowledgment.

- **Timely**
  - As we've discussed before, the closer the consequence is to the behavior, the more impactful the consequence will be. If your feedback is frequent, then it should also be timely. For it to be timely, it must be quick. And for it to be quick and still be effective, it must be precise!

EXAMPLE

When coaching Ms. Reid, the goal is to provide a quick interaction that won't keep her from her afterschool duties but allows ample time to note her past performance, note her current progress, and ask questions about what she feels is working and what she feels is not. This is the perfect opportunity for PIVOTal Feedback.

And that's essentially what you end up doing. You are able to observe and provide feedback without delay once the students are out of the classroom. You provide a positive feedback statement that is specific to her improvements in getting the students to attend, and you discuss the value of this improvement in her day, the ease of the activity given these improvements, and the impact on student outcomes.

WHAT HAPPENED?

After you provided the coaching interaction, what happened? Did the behavior change? Have there been any changes in results yet? What did the person do while you were coaching them? The goal is to get results, but to get results in the right way to build a positive school culture. This means that you need to monitor results, monitor the behaviors of the teacher that led to those results, and monitor the body language and interaction the teacher has as you're coaching them. If you're getting results and behavior change, but the teacher is avoiding eye contact with you or deflecting blame (or credit) to others during constructive and positive feedback conversations, this could be an indicator that the teacher doesn't like or see value in your coaching interactions. This could lead to the teacher doing just enough to avoid having those coaching interactions with you, which is not good. You want a teacher to not only see value in your coaching, but to see so much value that they actually start approaching you to talk about their performance because good things happen when they do.

EXAMPLE

Ms. Reid initially seems uncomfortable with your observations. She's not used to being regularly tracked, and historically, tracking performance meant she'd only hear from her previous administrator when something was wrong. But now she seems eager to discuss her progress after class. She smiles and actually approaches you in the hallway before you even get to her classroom to tell you about the progress she's made. And it seems to be working so far. Student test scores are increasing, and it appears that almost all students are attending prior to the math lessons.

WHY DON'T ... ?

"Why don't you try the changes we discussed during your next class, and I'll check back to see how things went?" "Why don't you send me an email once the changes are made to your curriculum and let me know if there's anything I can do to help?" The Deliberate Coaching process doesn't stop because teacher performance doesn't stop. Shaping behavior into long-lasting habits takes time. Educators always need to be recognized for the good work they're doing, and they need support to help them make tweaks here and there to maximize performance.

EXAMPLE

Ms. Reid is making real progress, so your next steps are to continue to reinforce her progress and emphasize how her improvements are impacting her day-to-day work and student outcomes. Asking questions such as "How is this making your day better?" could get her talking about natural reinforcers!

The 8 Ws form should be one of the first tools you use, as it's designed to both prompt your coaching behaviors and track progress. When coaching coaches to use this form, remember that, as with any new behavior, you want to reinforce its use, so start small. First, simply reinforce bringing the form and filling it out, in whatever way the coach likes. Then you can start shaping up the quality and constancy of the information in the form. But using it is imperative. If they don't, they'll be leading without any order or structure.

Coaching With Questioning

Since we don't want teachers to always rely on coaches telling them what to do and why doing it is important, what kinds of questions should be asked? There are actually quite a few books dedicated to this very subject. However, Deliberate Coaching is primarily concerned with two things: behavior and results.

Having a good "ask" through coaching questions is an effective method for engaging the teacher in their own performance improvement. Questioning allows the teacher to reflect on the link between their behavior and outcomes; moreover, it creates many more opportunities for the coach to gather information that can be used to guide coaching interactions.

The function of questioning, therefore, should be to help the educator recognize the impact of their behavior on the desired result and to provide the coach with an opportunity to assess and give feedback. As such, we recommend that Deliberate Coaches seek precise behavioral responses that get educators to self-identify:

- their own behavior (good and bad)
- the results of their behavior
- why the behavior is occurring
- what to do instead
- why the behavior is important (particularly in terms of the impact on consequences that matter to them personally)

Good coaches ask great questions to get educators in touch with naturally occurring positive reinforcement. They do this by helping educators look for even the smallest positive changes in their environment that result from their behavior. When this happens, educators will continue attempting a task, skill, or process, even in the absence of the coach.

Raising good questions is not new to educators or to those coaching. An art that requires cultivated and practiced knowledge (Cavanaugh & Warwick, 2001), it is fundamental to stimulating student thinking (Aschner, 1961). Posing questions pertinent to a specific scope of knowledge will facilitate the learning process. In the field of language learning and teaching, the importance of questioning as a teaching and learning strategy is well documented (Albergaria-Almeida, 2010; Chin, 2007; Chin & Osborne, 2008; Graesser & Olde, 2003; Roth, 1996). Students' level of engagement largely depends on the questions formulated by the teachers in the classroom that prompt and guide thinking (Wilen, 1991) and on the questions generated by students themselves in the process of learning and teaching (Albergaria-Almeida, 2010, 2012).

This concept applies to adults as well; many coaching experts (e.g., Knight, 2009; Maxwell, 2014; Souza, 2015; Wilson, 2007) advocate questioning be used to allow the coachee to reflect on their practice. This also allows both the coach and coachee to collaboratively gain new insights and better inform future practice. Indeed, there is an ongoing and still-developing focus among researchers on questioning strategies, as they're an indispensable element in developing, expanding, and challenging thinking (Klem & Connell, 2004; Marzano et al., 2001; Miciano, 2004).

The use of questions is a topic that is addressed by outlets such as the *Harvard Business Review* and the International Coaching Federation and is an area that continues to be researched (e.g., Patterson et al. 2022; Su, 2014; Wallis, 2016).

Questioning to Avoid Prompt Dependency

Say it's late in the afternoon and you are making the drive to see your friend an hour north. Though you've visited him many times over the past year since he's moved to the area, you plug his address into the GPS as you always do, so you can put your mind, along with the car, on cruise control. You sit back, turn the radio on, and sing along to kill time.

About 50 minutes into the drive, the unthinkable happens. After going haywire for a moment, the GPS shuts down, never to grace the interior of your car with its colorful screen again. Before you have a chance to mourn the death of your former electronic companion, you realize that you have no idea where you are! Though you've made the trip dozens of times, your now defunct electronic buddy has always guided your every move. As a result, you don't recognize the landmarks or even the exit where you normally get off, as they all look similar.

In the story above, you've become a victim to the phenomenon known as *prompt dependency* (Cooper et al., 2020). In short, prompt dependency occurs when somebody regularly needs prompting in the form of instruction, explanations, illustrations, or even gestures to complete a task. Prompting in and of itself isn't a bad thing—it's actually critical to teaching and learning during early stages of skill acquisition. However, when people are always told what to do during training or coaching as opposed to responding to natural cues, they tend to become dependent on the trainer or coach. This is likely because they haven't had to exercise important skills such as discriminating, problem-solving, or decision-making to accomplish the task.

To avoid prompt dependency, behavior science recommends a powerful yet simple procedure: *fading*. Fading occurs when prompts are systematically reduced. While the term might seem foreign, you've done it and had it done to you often throughout your life. For example, before you learned to walk independently, your parent might have first held both hands, then one hand, and then let you hold a finger. This systematic fading of physical guidance occurred until you acquired the correct balance and movement and no longer needed the additional help. The same thing occurred when you learned to discriminate between colors. Your mom might have said to you, "That's reeeedddddd. What color is it?" And you responded, "Red!" After some repetition, you were eventually able to respond correctly, without the precorrection. Your mom simply used a question—"What color is it?"—to evoke the appropriate response, "Red!" While this kind of questioning is commonly used on children, many adults don't understand how powerful it actually is.

Planning, Tracking, and Follow-Up

Remember that Deliberate Coaching is an ongoing process. The 8 Ws of Deliberate Coaching are designed to keep you focused on a precise target based on the needs of the educator being coached. Following these eight areas allows you to plan your coaching, track how things are going, and use this information to keep the progress coming strategically and steadily. The clarity of how you communicate about your coaching is critical, both for personal growth as a leader and for transparency with those you're working for and with. Your goal is to make the coaching process valuable to your direct reports by helping them with things that matter. You do this by being precise, purposeful, and systematic. And you build a positive school culture by keeping this going so that you and the other coaches make the lives of all educators and staff members better through your support. In turn, they will make the lives of the students better, which is our ultimate goal.

Tracking can simply mean providing a brief description of what you're seeing. You might even just create a couple of boxes (see Figure 10.2) as a way to quickly note progress on your pinpoints and linked results. Keeping these together helps align the two, making sure your focus remains on both the specific teacher behavior and the reason why that behavior is important (the outcome).

You might also wish to consider a more formal way of tracking teacher performance via the related results. You want your data to be as objective as possible, but we understand that measurement can be time-consuming. So do what you can, but the more data you gather, the better! The information you gather should tell a story about how things are going—what's working and what's not working. The more precise these data are, the more buy-in you'll get. If you make subjective judgements about behavior, then you might get into a battle of what you think is going on versus what the teacher says is going on. That isn't a battle you want to be in, and there's no way for either person to win it. You can prevent this issue by getting more precise data. Cut off potential confrontations through objectivity. Try observing and counting the behaviors. Or have the teacher keep a tally.

For instance, if you want to know how many times a teacher asked for engagement during a lesson, try to tally this up. If you want to know how many of those

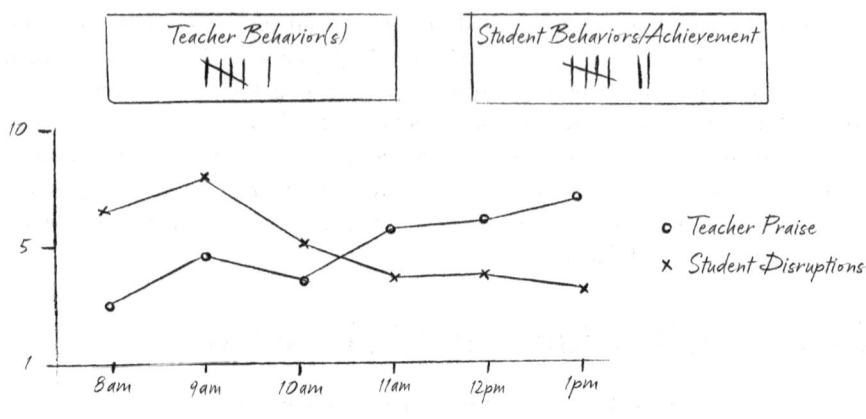

Figure 10.2. *Tracking behavior and results is critical to ensure you're working on the right behaviors and those behaviors are having an impact. However, you do not need a complex measurement system. Sometimes, it can be as easy as some quick tally marks and accompanying simple graphs to track progress.*

answers were correct, mark it and tally a percentage. You can create a simple form like the boxes to keep track of your tallies, and something like the simple graph options depicted in Figure 10.2 as a coaching tool to help you and your coachee observe trends.

SEEING PROGRESS TO REINFORCE PROGRESS

According to a popular saying, you don't know where you're going until you know where you've been. Coaches can't coach if they don't know what they're coaching and how it's impacting performance. That's how we show the value of what we do—we need to track performance to see it. Teachers show value the same way, with others judging it based on the performance of their students. But vice principals, principals, and other school coaches are in a position to get only a snapshot of what the teachers are doing to get that performance.

A survey conducted by the ROI Institute sought the opinions of executives on training and development (Phillips, 2010). Targeting CEOs of Fortune 500 organizations and large private firms, it provided a unique glimpse into the mindset of the individuals responsible for funding their training efforts. The survey covered various topics reported by their training and development departments, such as how high employees rate the training, the impact training has on top business metrics, training efficiency, return on investment, and how much employees learn as a result of the training.

According to the CEOs, the two factors their training departments considered most significant were the number of employees trained in a given year and the training efficiency (e.g., cost per hour of learning). Coincidentally, these were also at the *bottom* of the CEOs' list. Instead, they believed that the *impact* of training on key business metrics and return on investment (cost/benefit) were among the most important factors. And what reports did the CEOs report receiving the least? You guessed it—impact metrics and ROI.

Why are these findings so important? They show that it's easy to ignore the message we're trying to send with our outcomes. As a coach, do you know what the school administration looks at to see the value in what you're doing? As a school administrator, do you know what the school board looks at to see the value in what you do? What about parents and key stakeholders in the community? Our school systems are just that: *systems*. Specifically, they are *performance systems*, requiring resources and support to guide the individuals responsible for producing results.

Noted in the CEO survey was that the chief learning officer—the head of training and development—is approximately three levels below the CEO, leaving a sizeable hill to climb to provide and maintain support. How far removed are you from the key decision-makers in your school district, those individuals responsible for giving you the resources and flexibility you need to be a successful coach? Sustainable support and growth in our school systems comes when we show that our work adds value that's greater than the cost required to attain meaningful results. Specifically, we must show value in a precise, objective, and replicable way.

THE COACHING TELEPHONE GAME

Consider this scenario. You're walking down the hallway to pick your child up from daycare when you see a group of kids circling their teacher in anticipation. The teacher whispers something in the ear of the child closest to her and asks her to pass it along. The girl whispers in the ear of the boy next to her, who then does the same to the boy next to him, and so on until the final child gets the message. At this point the teacher asks the last child to recite the message: "No planks hidden in the monsoon," the child says with confidence. "And Olivia, what was the message I gave you?" Olivia laughs and says, "Snowflakes will fall softly soon."

What does the telephone game have to do with your coaching system? Let me paint you a picture of the *coaching* telephone game:

- School decision-makers want to invest in people and systems that add value to the school.

- To get an idea of what has value and what doesn't, they'll either be *told* or be *shown*.

- If you don't report directly to these decision-makers as a coach, this means you have to find a way to relay the information up the chain of command. If you rely on subjective descriptions and verbal reports, you're leaving your impact up to interpretation. By the time it gets to the school decision-makers, they might be wondering how "planks and monsoons" are worth the school's continued support.

WHAT TO DO INSTEAD

The Deliberate Coaching process is designed to allow small tweaks in the way you affect teacher performance. However, this process still requires support from senior school leaders. These key individuals are essential to ensuring that your impact is maximized for as long as you're with the school. Think of the behaviors you want from these individuals. Perhaps you want them to sign off on assigning or hiring more coaches, to buy more materials, or to follow through on suggestions you made about how to properly utilize your coaching recommendations. If these behaviors are to occur, then they, like all other behaviors, need motivation—something you can reinforce by following a few suggestions:

1. Train for Impact

To demonstrate value up the chain of command, you must control your message. You do this by pinpointing critical metrics in a way that is objective. Precision is key here. This enables you to clearly present your value in the form of concise, objective impact data and reports. Think of the numbers that come up in discussions of how much your school is progressing or how strong a particular teacher is. These could be things like standardized test scores, the cost of student progress, expenses, or teacher retention. These impact data will not only help you control your value message, but they'll also help guide your coaching efforts and allow you to continuously improve your coaching systems, staying one step ahead by tracking the most critical metrics.

Identifying critical impact results is important, but getting results for things such as performance evaluations, turnover data, and school costs can take time—valuable time that you could spend making appropriate adjustments and fine-tuning your coaching to ensure that those delayed results will show something positive. You need more leading indicators of success, things such as teachers' day-to-day attendance and timeliness, curriculum adjustments, student progress, or teacher feedback surveys. You find these by starting with the lagging metrics that show what is important to the school (like graduation rates and teacher turnover) yet are taken too infrequently to use as a primary way to show what is working and what is not. From these lagging indicators, you can then identify the leading indicators that you have a constant and direct impact on. This frequent tracking allows you to identify strengths of your coaching and areas that need improvement. By impacting these leading indicators, you'll be able to proactively ensure that you'll have a positive impact on the infrequent results that are most critical to the school and the school decision-makers.

2. Tracking the Behavior Leads to Result Alignment

As you track progress, your goal is to showcase how the critical teacher behaviors you're coaching are impacting results that matter. As an individual responsible for teacher development, your job is to build these behaviors. If you build the wrong behaviors, you risk spending time and money on skills that fail to have an effect on anything, ultimately producing unhappy school administrators who might not support your next initiative.

Depending on your role, you might be responsible for your school's training and coaching systems—or at least have access to the person who does. Regardless, start off by identifying the teacher behaviors that will have the greatest impact on the critical measures of success you're tracking for the school, classroom, and students. If you don't know what these behaviors are, then ask. Better to find a group of exemplars and find out what they do to be successful than guess or try to remember what you used to do prior to your promotion. These teachers are responsible for the work, whereas their supervisors are responsible for *influencing* the work.

You can make the greatest impact on your key performance outcomes by prioritizing those doing the work. So if you can identify the right teacher behaviors to coach—those that have a clear link to critical results—you'll be ready to develop a plan to target those behaviors. Without truly knowing what those behaviors are, you're leaving your coaching (and potentially your job) up to chance ... and that's not good science!

If you're responsible for training and coaching teachers, then you might also be responsible for training other coaches and leaders. For this group, continue aligning critical behaviors and results to identify the key leadership behaviors that have the greatest impact on the performance of the teachers. Again, if you don't know what those behaviors are, find the exemplars and ask them to describe how they've achieved success as a leader.

But a word of caution from your friendly neighborhood behavior analysts. It's imperative that you don't just find people who get the best results. A bad leader can get good results in the wrong way, and you don't want to create an army of these individuals. It's easy to fall into the trap of assuming that just because Mr. Cohen has been a vice principal for 20 years and seems to get the results the school is looking for, his do-it-or-else approach to leadership is something to replicate. In the same regard, you also don't necessarily want to find the manager who is the most liked. You want to create a team of leaders who know how to lead, not just how to be your friend. The sweet spot is the leader who gets results in the right way. These are the supportive and proactive leaders who lead by positively reinforcing good performance. Find these individuals and ask them to describe what they do to be effective. Just as you did when identifying your results, you're looking for specific pinpointed behaviors that can be measured. Ask the leader questions like "What does *leading by example* look like?" to really find the behaviors. Vague descriptions of behavior that lack precision and focus will produce vague coaching systems that lack precision and focus.

3. Demonstrate Value With Measurement

One of the most powerful tools in your science toolbox is your measurement system. Good data help take away the murky bias of subjectivity and allow for a breath of objective fresh air. Here are a few stories that might help showcase the value of tracking and aligning your impact.

MR. ABBOUD'S STORY

Mr. Abboud is responsible for professional development at his school and has heard great things about his training and coaching. After the number of positive reports decreases, he decides that his systems need an overhaul. Now he's getting more complaints than ever. He's always been against relying on data to show that his leadership works, confident in his process and impact. But his confidence is starting to waver.

WHAT MR. ABBOUD SHOULD DO

Measurements and data are there to tell a story. When it comes to measuring the impact of your coaching system, they're here to tell *your* story. They should be embraced for their reinforcement and decision-making potential, not rejected.

Mr. Abboud was relying on the coaching telephone game to evaluate his coaching, then using this information to guide serious changes in his leadership system. If you're not tracking progress, this approach can result in costly errors. You might continue to replicate an ineffective training only to later find out that your training is missing the mark. Even worse, you could stop or change an *effective* coaching system because of anecdotal or subjective reports that have nothing to do with your true training impact.

MS. LEE'S STORY

Ms. Lee is a principal who's also failing to see meaningful change. Unlike Mr. Abboud, she's been tracking impact metrics. Historically, her adherence to the annual budget targets, coupled with her school's pristine graduation rate, has shown the kind of results that make her school a highly regarded asset for the district. But recent layoffs, budget cuts, and unexpected policy changes have left her overwhelmed, and her impact metrics are taking a turn for the worse.

WHAT MS. LEE SHOULD DO

Ms. Lee is losing control of her impact. There are always going to be lagging metrics caused by variables that you and your teachers have little to no influence over. You can't control district-level policy changes that negatively affect your day-to-day operations, and you can't prevent layoffs and budget cuts that can affect classroom ratios, materials, and your school's overall fiscal and academic strength. However, you *can* control how you choose to measure and evaluate your impact as a leader and the impact of those coaches working with you. Remember that leading indicators are designed to help you not only continuously refine your coaching, but also present an accurate message to the district on the impact and value of these efforts. Ms. Lee hasn't been tracking leading indicators of her coaching impact, instead she has been relying solely on lagging metrics—specifically the lagging metrics she has little control over. Without leading metrics, her message is intertwined with lagging results that don't fully capture her impact. Avoid relying on impact metrics that are outside of the control of your teachers and staff.

MR. RUIZ'S STORY

Mr. Ruiz seems to be a step ahead of the rest. As a vice principal, he discusses how he's reduced teacher tardiness and made timely curriculum improvements over the past few months, and he reports that his feedback from teachers has been consistently positive. His troubles come when his principal pulls him aside to discuss the impact of his coaching. It turns out that standardized test scores have been declining and the number of student suspensions is up. The district is requesting that his coaching system be revised to address these issues.

WHAT MR. RUIZ SHOULD DO

Unlike Ms. Lee, Mr. Ruiz has tracked both leading and lagging metrics. But while he's used his leading metrics to help his teachers and evaluate his own impact, he hasn't shared those metrics with his principal. He's also failed to track other critical outcomes to ensure that they aren't being neglected. Allowing decision-makers to act based on incomplete information can be costly.

Let your data tell the story so you can see what's working and why, then expand the value of these effective systems. Help your school find the true cause of a problem so it doesn't stop the one system that actually works. When your leading indicators are well pinpointed and linked to the right behaviors, you can demonstrate that the cause of subpar lagging results might not be poor training or coaching. If you don't have a seat at that table when discussing your impact, try to ensure that your data do.

Showcasing the link between key results and the behaviors you're coaching is just as important as the effectiveness of the coaching itself. Long-term administrative support depends on our training and coaching efforts remaining on the radar of key decision-makers. By understanding the expected impact of your leadership system, you can work to find the critical behaviors that should be guiding your coaching efforts. If you don't know the behaviors that lead to valued impact, you can't reliably build them. Our goal is valued results that can be reliably replicated, effectively disseminated, and clearly communicated.

Key Takeaways

- The 8 Ws of Deliberate Coaching is a quick and highly effective tool that acts both as a guide for planning your coaching and as a way to report how your coaching interactions went.

- Who? Individualization is key; coach one teacher at a time, recognizing unique motivations and challenges.

- Why? Purpose matters; your coaching should be a strategic effort to foster growth and address obstacles.

- Which? Clarity is crucial; pinpoint specific results and behaviors to target for improvement.

- When? Timing is critical; identify when targeted behaviors occur for impactful coaching.

- Where? Environment is influential; select a conducive setting for coaching to enhance comfort and receptiveness.

- What? Action drives progress; focus on what needs to be done, from observation to providing feedback.

- What Happened? Observe outcomes; assess whether behaviors and results improve post-coaching.

- Why Don't ... ? Encourage actionable steps; suggest strategies for continuous improvement.

- PIVOT your feedback conversations to maximize the impact of your coaching by using precise, individualized, valued, ongoing, and timely feedback.

- Talking *at* someone closes a conversation. Using open-ended questions can engage people and help them become good observers of their behavior and the impact of their behavior, which are the critical components to building habits.

11

Precise Coaching

It is the mark of an educated person to search for the same kind of clarity in each topic to the extent that the nature of the matter accepts it.

—ARISTOTLE, *NICOMACHEAN ETHICS*

In performance improvement, pinpointing is the process of describing precisely what outcome is desired and precisely what behavior should occur to achieve that outcome (Daniels & Bailey, 2006). It may sound oversimplistic, but failure to accurately pinpoint both the outcome *and* the behavior is responsible for many of the performance issues that exist in organizations. Remember, all results require somebody doing something more, less, or differently. A focus on outcome data alone (e.g., student achievement) disregards the actual behavior required to achieve desired results.

As you know, one of the guiding principles of Deliberate Coaching is precision. And precision is at the root of pinpointing. It first requires a careful analysis of the desired results as well as the behaviors required to achieve that outcome. Not just any results or behaviors, mind you, but *precise* results and specific, observable behaviors. Simply put, you're asking where we want to go and how we want to get there. But be careful, sometimes it's not as easy as it sounds.

Imagine vacationing in Orlando with your family. While eating dinner one evening, you overhear a couple talking about their trip to a beautiful, secluded beach in Florida. It sounds like an amazing place, so you lean over, apologize for eavesdropping, and politely ask where the beach is. The couple laughs. They share a few more tantalizing facts about the beach and tourist attractions in the surrounding area, then give you the name of the town. "A trip to Four Candles would be a great experience for any family," they say.

The next morning, you excitedly pack your bags, load up the rented minivan, and fumble with the rental's GPS until you're able to enter Four Candles as a destination. After 3 hours of traveling southeast, following the precise directions provided by the GPS, you arrive at Four Candles. While the town appears to be pleasant enough, it isn't at all what was described by the couple. In fact, when you make your way to the beach, you're surprised to find it isn't secluded at all, but rather overcrowded.

Confused, you park the car and walk up to the local lifeguard to inquire about the secluded beach you heard about. The lifeguard chuckles. "This is the only beach in Four Candles, and it's always packed!"

"Where's the Four Candles trolley tour we heard about?"

"Trolley tour?" the lifeguard repeats, confused.

"Yes, we were told by a nice couple in Orlando that Four Candles had a secluded beach and a town with quite a few tourist attractions."

The lifeguard takes this in. "I think I know what happened. You've confused Four Candles with Fork Handles, which is a town with tourist attractions and a secluded beach. The only problem is, it's about five hours away, on the *west coast* of Florida."

Though you followed all of the GPS's directions precisely, you didn't achieve the desired outcome because the destination and the precise behaviors or "path" to get there wasn't pinpointed. Similarly, while the ultimate destination in education is student achievement, you must take care to identify long- *and* short-term goals that will provide a clear path that will eventually lead to desired growth. This requires precise and purposeful planning—you must figure out exactly which behaviors and goals will eventually lead to the kind of achievement you desire.

As a coach, determining the exact result is critical. It's not unusual for folks to jump right in with the good intention of helping a teacher in need. If you ever feel like doing this, remember the story above. If you'd done a little research on the internet prior to getting on the road, you and your family would have reached the *correct* destination in a reasonable amount of time. If you were traveling the old-fashioned way using a paper map, you would have likely identified some destinations along the way as benchmarks to let you know you were headed in the right direction as opposed to arriving and realizing you were in the wrong spot. These initial destinations are akin to short-term goals. They aren't the desired outcome, but they're leading indicators of success that let you know you're on the right path. These leading indicators are metrics that are taken frequently enough that you can use them to continuously track how well your coaching is doing.

> In a similar scenario, imagine being provided with the correct destination with handwritten directions (your GPS is not working), but you fail to reach the correct location as a result of a simple mistake. Specifically, somewhere along the trip you made a right turn when you should have made a left. In this case, the pinpointed behaviors outlined in the directions (e.g., drive north there, turn left here, etc.) were off. As a result of one small error, you didn't reach your destination.

The point is that pinpointing necessitates careful thought. It's not something that should be rushed into so you can jump in and do something. It requires a clear understanding of precisely where you want to go and precisely how you are going to get there. Let's take a look at the importance of precision in the school setting.

Precision in Educational Leadership: Lessons Learned

> Imagine Ms. Nowak, an experienced principal, planning a schoolwide initiative to improve reading comprehension scores. She gathers her team and announces the goal: "We need to boost our reading scores." Enthusiastic, but without specific guidance, personnel in each department start working independently, employing various strategies. Three months in, Ms. Nowak reviews the progress and finds minimal improvement. While her goal was clear, the lack of precise, actionable steps led to a disjointed effort and suboptimal results.

> Now consider a different approach. Ms. Nowak, informed by her previous experience, decides to implement a more precise plan. She starts with a detailed analysis of the current reading levels across grades and identifies key areas needing improvement. She then works with her team to develop specific, observable behaviors and strategies tailored to each grade's needs. For example, for lower grades, the focus might be on phonemic awareness exercises, while upper grades might concentrate on critical analysis of texts. She also sets clear benchmarks for assessing progress, such as monthly reading assessments and classroom observations.

> This time, the outcome is different. The staff, with clear directions and specific behaviors to focus on, works coherently toward the common goal. Regular assessments and observations provide ongoing feedback, allowing for adjustments along the way. The result is a significant improvement in reading comprehension scores across all grades.

This scenario highlights the critical role of precision in educational leadership. Just as in the story of the vacation gone awry due to imprecise information, in education, the path to achieving goals requires careful planning, clear communication of specific actions, and regular monitoring. Without precise and purposeful planning that outlines the exact behaviors and goals needed, even well-intentioned efforts can lead to misdirection and missed opportunities. By pinpointing the desired outcomes and the specific paths to achieve them, educational leaders can effectively guide their schools toward meaningful and measurable success.

IMPACT Pinpoints

When it comes to performance improvement, especially in educational settings, the concept of precision through pinpointing is pivotal. Recall that in Deliberate Coaching, precision is key. The IMPACT (Gavoni & Costa, 2023) approach, initially applied to goal setting in Chapter 5, can also be adapted to enhance the pinpointing process. Here's how we can redefine pinpointing using the IMPACT acronym:

Individualized: Pinpointing must be tailored to the specific context and needs of the individual or group involved. It's about identifying outcomes and behaviors that are directly relevant to the roles and responsibilities of those involved. For example, if a teacher struggles with classroom engagement, an individualized pinpoint might focus on specific strategies to increase student participation.

Manageable: The behaviors identified in the pinpointing process need to be achievable given the individual's current skill set and resources. Overly ambitious or complex behaviors can render the pinpoint unmanageable, leading to frustration and suboptimal outcomes.

Positively Motivating: The pinpointed behaviors and outcomes should be motivating for the individual. This involves ensuring that the goals are not just important in principle but also perceived as valuable and rewarding by those individuals expected to achieve them.

Aligned: Each pinpointed behavior should align closely with the ultimate outcome. It is critical to ensure that every identified action is a step toward the desired result and that each action contributes directly to achieving that result. This alignment ensures that effort is focused and effective.

Connected: The desired outcome and the behaviors to achieve it should be connected in a clear, logical manner through accomplishments. This connection is crucial for understanding how specific actions will lead to the attainment of the desired results.

Trackable: Both the outcomes and behaviors in pinpointing must be measurable. If you can't measure it, you can't manage it. This means both outcomes and behaviors must be defined in clear, observable terms, and the methods for progress tracking and evaluation must be clearly established.

By applying the IMPACT framework to the process of pinpointing, Deliberate Coaching becomes more precise, purposeful, and systematically aligned with the actual needs and capabilities of educators and staff. This refined approach to pinpointing not only clarifies what needs to be done but also significantly enhances the likelihood of achieving the desired performance improvements in an educational setting.

Consider using the IMPACT Pinpoints Checklist (Figure 11.1) to build this process using your own results and your own behavior pinpoints. This checklist will help you walk through the process of checking off the IMPACT criteria while aligning results and behavior.

Teacher Name: _____ Date: _____

IMPACT Pinpoints Checklist

Individual	Manageable	Positively Motivating	Aligned	Connected	Trackable
Is the pinpoint specific to the context and needs of the individual or group involved?	Are the behaviors achievable given the individual's current skill set and resources?	Is the pinpoint a behavior that can lead to meaningful change?	Are the behaviors and results aligned?	Are results connected to behaviors through accomplishments?	Are both the pinpointed behaviors and results measurable?

Pinpointed result:

Why is the result socially valid?

Pinpointed behaviors:

1.
2.
3.
4.

What accomplishments will be used as leading indicators to measure progress?

Equipment, process, training, or support needed to achieve this result:

Figure 11.1. Clear, objective pinpoints, for both behaviors and linked results, are critical to communicate measurable expectations and ensure alignment with members of your team. The IMPACT Pinpoints Checklist is a way to build quick and easy pinpoints.

CHOOSING THE RIGHT PINPOINTS

Imagine walking into a teacher's class for the first time and encountering chaos. Some students are talking during the lesson, two students appear to be sleeping, and a couple of students continue to ignore the teacher after they are asked to have a seat. So where do you start?

Three concepts will enable you to determine pinpoints: Quick Wins, stages of concern, and behavioral cusps.

Quick Wins

If you've just begun working with a teacher, we would suggest starting with a "Quick Win" (Gavoni & Costa, 2023). A Quick Win entails helping the teacher complete relatively easy tasks before you make requests that might be more challenging; moreover, a Quick Win requires pinpointing a result that will be meaningful to the teacher and that will make an easily observable difference. While it might not have an immediate and direct impact on the bottom line of student achievement, it's ultimately aligned with that goal (e.g., increasing on-task behavior is a prerequisite for learning).

A Quick Win in the example above might be to collaborate with the teacher to develop a plan that addresses behavior issues during morning arrival. Depending on the need, the Quick Win might be something as simple as having assigned seats for students and a question written on the board as a primer for the morning lesson. If this strategy results in a visible improvement in student behavior during morning arrival, you've just achieved a Quick Win, and the teacher is more likely to work with you in the future.

A major benefit of choosing a Quick Win is that it establishes you as a reinforcer: Because you helped the teacher implement a relatively simple strategy that resulted in a meaningful outcome, they'll be more likely to seek you out for assistance in the future. And as we suggested earlier, they'll also be more likely to attempt progressively more complex tasks in the future.

Stages of Concern

When pinpointing, it's critical that the teacher be involved in the process. While they may not always be aware of which behaviors lead to the results they want, they know *what* results they want. Determining which short- and long-term outcomes are important to them ensures that the pinpoints are relevant.

Research related to stages of concern supports the idea of choosing outcomes that are immediately meaningful to the teacher. Developed in the 1970s and 1980s by investigators at the University of Texas at Austin, the stages of concern are broken down into seven categories that provide a framework of what's important to a teacher as they progress in learning and applying a skill or "innovation" within the classroom (George et al., 2006). According to the researchers, teachers are more likely to have self-focused concerns during early stages of change; however, as they become more proficient in their job, they begin shifting their focus to the broader impact of their performance. Figure 11.2 outlines these stages of concern.

Stages of Concern			
I M P A C T	6	Refocusing	The individual focuses on exploring ways to reap more universal benefits from the innovation, including the possibility of making major changes to it or replacing it with a more powerful alternative.
	5	Collaboration	The individual focuses on coordinating and cooperating with others regarding use of the innovation.
	4	Consequence	The individual focuses on the innovation's impact on students in their immediate sphere of influence. Considerations include the relevance of the innovation for students; the evaluation of student outcomes, including performance and competencies; and the changes needed to improve student outcomes.
T A S K	3	Management	The individual focuses on the processes and tasks of using the innovation and the best use of information and resources. Issues related to efficiency, organizing, managing, and scheduling dominate.
S E L F	2	Personal	The individual is uncertain about the demands of the innovation, their adequacy to meet those demands, and/or their role with the innovation. The individual is analyzing their relationship to the reward structure of the organization, determining their part in decision-making, and considering potential conflicts with existing structures or personal commitment. Concerns also might involve the financial or status implications of the program for the individual and colleagues.
	1	Informational	The individual indicates a general awareness of the innovation and interest in learning more details about it. The individual does not seem to be worried about themselves in relation to the innovation. Any interest is in the impersonal, substantive aspects of the innovation, such as its general characteristics, effects, and requirements for use.
	0	Unconcerned	The individual indicates little concern about or involvement with the innovation.

Figure 11.2. *The stages of concern highlight how the value of different outcomes can change throughout a teacher's career. Understanding what outcomes we value and are valued by those we work with helps provide a line of sight to build the behaviors that lead to those outcomes, while also not neglecting other critical outcomes. Adapted from George et al. (2006).*

In behavioral terms, this might be considered a reinforcement hierarchy (Merrett & Musgrove, 1982), as it identifies what's actually important to the teacher at a given time. A new teacher's focus might move from basic issues and themselves to more complex issues and student outcomes as they become more seasoned. The scale in Figure 11.3 is a simplified, useful tool regarding the stages of concern.

Stages of Concern		Expressions of Concern
"Impact"	6	I have some ideas about something that would work even better.
	5	I would like to coordinate my effort with others, to maximize the innovation's effect.
	4	I need to know how my using it is affecting my students.
"Task"	3	I seem to be spending all my time getting materials ready.
"Self"	2	I need to know how using it will affect me.
	1	I would like to know more about it.
"Unconcerned"	0	I am not concerned about it.

Figure 11.3. Teacher self-assessment version. Adapted from George et al. (2006).

Understanding what's relevant or reinforcing to the teacher is paramount. While you might want to dive right in and help a new teacher ask more complex questions in the classroom as a pinpoint, for example, he might initially be more concerned with how he'll be evaluated, where he should clock in, what he should bring to meetings, and how he should appropriately request time off when he's sick. Or, as in the example above, reducing misbehavior that's driving the teacher crazy would likely be an initial concern that immediately benefits "self" but is also clearly linked to the future goals of student achievement.

BEHAVIORAL CUSPS

As we discussed in Chapter 7, behavioral cusps offer another opportunity to find valued pinpoints to target with your coachee. According to Rosales-Ruiz and Baer (1997), a behavioral cusp is "a behavior change that has consequences for the organism beyond the change itself, some of which may be considered important" (p. 537). For example, consider the young child who learns to crawl. Where they may have initially been confined to what reinforcement was within their immediate reach, they now can access a variety of reinforcement (e.g., getting to siblings, playing with the dog, retrieving their favorite toy from the toy box) that was previously unavailable. As a result, other behaviors might begin to take shape such as social play, walking, or further developing fine motor skills.

The concept of a cusp encourages you to consider the future implications of a pinpointed behavior rather than focusing solely on the immediate impact. Let's take a look at the potential impact of pinpointing a behavioral cusp for teachers and leaders.

Behavioral Cusp Example for Teachers

Consider the skill of differentiated instruction. When a teacher learns and effectively implements differentiated instruction strategies, they have access to a range of positive outcomes not only for their teaching methods but also for student learning. Here are a few:

- Student understanding improves because the teacher is better able to cater to diverse learning styles and needs.

- Student participation increases because more students engage as the material becomes accessible to a broader range.

- Adaptive teaching skills continue to grow and develop as the teacher becomes more adept at adjusting lessons dynamically.

- Classroom management is easier because students are more engaged, leading to fewer behavioral issues.

- Teacher confidence increases because the teacher sees the impact of their teaching abilities with the students they're teaching.

- Student learning is enhanced and improved because the teacher is better able to provide more tailored feedback to students and to offer personalized student feedback.

Behavioral Cusp Example for School Leaders

For school leaders, the development of effective communication strategies can be a behavioral cusp. Mastering this skill can have significant, wide-reaching effects. These include:

- Staff morale and trust in leadership improve as a result of clear, transparent communication.

- Collaboration improves as a result of effective communication, which fosters a collaborative environment where ideas and strategies are freely shared.

- Conflict resolution improves as the leader is better equipped to handle disputes and issues that arise within the school.

- Community relations and school-community ties are strengthened as the leader is able to effectively communicate with parents and community members.

- Decisions become better informed due to open channels of communication that consider diverse perspectives.

- Overall organization health and leadership improve as a result of effective communication being the standard for communication throughout the school.

These examples underscore the importance of nurturing key behavioral cusps, which can catalyze profound and advantageous changes extending well beyond the initial behavior and can positively affect various facets of educational settings. In this context, several Deliberate Coaching strategies we've discussed can be seen as behavioral cusps. For instance, the practices of pinpointing, measuring, offering feedback, and providing positive reinforcement are all pivotal behaviors that can lead to broader, impactful changes.

Pinpointing: Pinpointing sets a clear direction and influences a range of subsequent actions and decisions by precisely identifying desired outcomes and required behaviors.

Measuring: Regular measurement of progress establishes accountability and continuous improvement and influences broader educational outcomes.

Providing Feedback: Constructive feedback guides and adjusts behaviors and leads to improved teaching practices and student learning experiences.

Delivering Positive Reinforcement: Positive reinforcement encourages repetition of effective behaviors and fosters an environment of success and motivation.

Each of these approaches, integral to Deliberate Coaching, serves as a behavioral cusp, opening doors to wider positive transformations in the educational realm.

Increasing Precision Through Task Analysis

Education, at its core, is about providing teachers with effective strategies to enhance learning. Instead of discussing the broad spectrum of pedagogical research, let's focus on a specific, practical approach: optimizing opportunities to respond.

For instance, consider the method of increasing opportunities for student responses. This is a fundamental strategy to maintain engagement and ensure active participation. Below is a breakdown of how this can be applied in a classroom setting without delving into specific research studies or organizations.

INCREASING OPPORTUNITIES TO RESPOND

1. **Initiation:** Start with an open-ended question to encourage thinking and engagement. This might involve presenting a problem that requires application of recently learned concepts, sparking curiosity and motivation.

2. **Wait Time:** After posing a question, pause for about 10 seconds. This gives students time to think and formulate their responses, fostering a deeper engagement with the material.

3. **Calling on Students:** Select students to respond in a manner that ensures all voices are heard. This can be random or strategically planned to involve those less inclined to participate.

4. **Variety in Responses:** Encourage different types of responses, such as verbal answers, written replies, or using digital tools. This diversity in response methods can cater to different learning styles and keep engagement high.

5. **Equity in Participation:** Make an effort to include every student over time. This involves balancing participation across the class, ensuring that all students have equal opportunities to contribute.

6. **Feedback:** Provide immediate and constructive feedback on responses. If a student's answer is correct, acknowledge and possibly expand on it. For partially correct answers, guide the student toward the full answer without discouragement. If an answer is incorrect, offer hints or scaffolded questions to help them arrive at the correct response on their own.

Such a task analysis serves as a guide for teaching strategies and a tool for self-reflection and continuous improvement through self-management and Deliberate Coaching of teaching practices. By breaking down the process into actionable steps, educators can more effectively self-assess and refine their approach to fostering an active learning environment, and leaders can do the same.

This information can be precisely described as part of the task analysis, used to train and coach opportunities to respond, used by teachers to remind them of what to do prior to instruction, or used by teachers as a self-assessment to determine how well they did after instruction. All of these marvelous performance-improvement tools blossom out of a good task analysis!

As the old saying goes, the devil is in the details. This is never more evident than when it comes to teaching. Sometimes, teachers do the right thing at the wrong time. Sometimes, they do the wrong thing at the right time. In both cases, precision is the issue. If the teacher calls on a student prior to asking the question in the task illustration above, the simple error in sequencing can have the unintended effect of disengaging the rest of the class. In other words, while the student who is called on must think about how to answer the question, other students might turn off their thinking caps as they now don't anticipate being called on. Asking the question first and then calling on a student, as indicated in the task analysis, is an effective strategy for engaging all of the students rather than just one.

Once again, precision is critical. When performing certain tasks, even one absent or incorrectly sequenced critical behavior (like making a right instead of a left) can have a ripple effect.

A final note regarding the number of steps or level of detail required of a task analysis. The key element to consider when developing it has to do with the skill level of the teacher as well as the teacher's prior experiences of learning or applying the skill. The greater the knowledge and skill, the less detail required.

Increasing Precision Through Checklists

THE USE OF CHECKLISTS—THE TALE OF TWO TEACHERS

> Ms. Lopez, a high school English teacher, recently attended a professional development training on integrating technology in classroom instruction. She is excited to implement what she learned but soon finds herself struggling. Despite her enthusiasm, she can't quite remember the sequence of steps to effectively incorporate the new software into her lesson plans. She attempts to follow through, but her lessons often end up disjointed and this frustrates both her and her students. Ms. Lopez's inability to recall the specific techniques from the training makes it challenging to apply them effectively in her classroom, so she reverts to her old teaching methods.
>
> Mr. Nguyen, also a high school teacher, attended the same training as Ms. Lopez. Recognizing the complexity of the new skills, he creates a checklist based on the training's content and breaks down the process of integrating technology into manageable steps. Each day, as he prepares and delivers his lessons, Mr. Nguyen refers to his checklist to ensure he covers each step. This methodical approach helps him implement the new techniques smoothly and leads to engaging and interactive lessons. His students are more involved, and the technology becomes an asset rather than a hindrance in his teaching. The checklist not only serves as a reminder but also as a tool for self-assessment so that Mr. Nguyen can track his progress and refine his skills continuously.

Most people are familiar with checklists. They're used for a variety of tasks including chores or shopping lists. Checklists are one of the most commonly used and effective antecedents, and they prompt the behaviors on the checklist and offer a way to track progress. While an unsophisticated means of remembering things, as can be observed from The Tale of Two Teachers vignette, a checklist is in fact a powerful tool for getting a variety of personal and professional tasks right. Used by a variety of professionals, including diligent surgeons, the development of the checklist is rooted in an unfortunate historical event (Gawande, 2010).

In 1935, after 5 years of development, the most sophisticated plane in aviation history sped down the runway, lifted off the pavement to begin ascent, then stalled and crashed in a fiery explosion that killed three people. Suspecting there was something wrong with the aircraft, investigators completed a thorough search. But to everybody's surprise, they found mechanical failure wasn't responsible for the crash. Human error was. As the plane became more sophisticated in its development, so did the techniques for flying it. Unfortunately, humans aren't infallible. Mistakes are a part of life. In most cases they can lead to learning, but in some cases, as in those related to flying or surgery, they can cost lives.

As a result of this unfortunate incident, the world witnessed the birth of the check-list, which is simply a task analysis that can be "checked" as steps happen. Checklists are powerful tools. They can be used for training, coaching, or even self-management.

In a training intervention, teachers can review the checklist as they practice the sequence of steps. In coaching, coaches can use this intervention to collect data and provide feedback to the teacher regarding the specific skills they got right and the ones they should deliberately focus on. And in self-management, Deliberate Coaches can assign the tool to teachers, who use it to remind themselves to implement a skill and assess their own performance afterward. In all cases, the checklist provides a metric that can help you continue to shape performance. To use it as a measure, you need only to divide the number of steps executed correctly by the total number of steps. For example, if a teacher gets 7 out of 10 steps right, they have a score of 70%.

Precision in your coaching will keep you and your coachee focused on behaviors and results that matter. If you're too vague, any coaching interactions could simply be white noise to the teacher you're working with. If you can't specify the behavior you're trying to affect and why it's important, you can't expect the teacher to change or to know what they're doing well.

Key Takeaways

- All results require somebody to do something more, less, or differently, and there must be alignment on what specific behaviors are being pinpointed.

- Precision is at the root of pinpointing and requires a careful analysis of the desired results as well as the behaviors required to achieve that outcome.

- Evaluate the precision of your pinpoints by ensuring the behavior is **I**ndividualized, **M**anageable, **P**ositively motivating, **A**ligned, **C**onnected, and **T**rackable.

- When prioritizing which pinpoints to address, consider the impact on the performer, the impact on the student, and alignment with valued results—and look for Quick Wins.

- Task Analysis:

  - Task analysis offers a structured approach for educators to understand and implement various teaching strategies. By breaking down tasks into manageable components, educators can more easily learn, apply, and refine these strategies in their teaching practices.

  - Educators can use task analysis as a self-assessment tool to evaluate their performance. This reflective practice allows for self-correction and improvement, ensuring that educators continually enhance their teaching methods and student engagement strategies.

  - Precision in teaching practices is critical. A well-conducted task analysis helps ensure that each step of a teaching strategy is executed correctly and in the right order. This precision prevents confusion and disengagement, maximizing student participation and learning outcomes.

  - Task analysis is versatile and can be tailored to the specific needs and skill levels of educators. Novice teachers may benefit from detailed breakdowns, while experienced educators might use task analyses to refine and adjust their techniques for even better outcomes.

  - The task analysis serves as an observation and feedback for the Deliberate Coach, facilitating targeted feedback and actionable insights for educators.

12

Purposeful Coaching

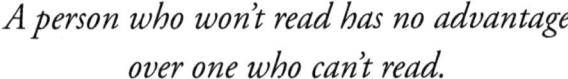

*A person who won't read has no advantage
over one who can't read.*

—Unknown

Long-lasting and efficient behavior change requires a planned, strategic approach. Behavior is shaped by celebrating small successes and encouraging progress until the behavior is at a goal level. This powerful scientific process is not random; leaders must breathe life into that fire to keep it going, or the behavior will be extinguished.

Coaching must be purposeful, or intentionally designed with a focus on developing a few skills at a time through much consistent repetition and feedback. This is especially important in an educational setting where students often have summers off with limited learning opportunities. If the skills taught to our students are not embedded into day-to-day activities (e.g., home skills, games, sports) or taught to fast and accurate fluency standards during the school year, then many of the skills won't last through summer. This means that an already short academic year gets shorter, using those first few weeks to get students caught up and back in the swing of learning. Add in breaks, testing, and sick days, and you can see why the time educators have is so precious.

To take advantage of that time, you have to be precise with your interactions— you don't have time for multiple 30-minute coaching conversations with each teacher each day. So how do you choose a course of action? How do you know whether you should be modeling the correct behavior for the teacher, providing feedback, prompting a teacher to check the curriculum before a lesson, or any number of other potential coaching solutions? Going with the wrong solution isn't the end of the world, but it will waste the time we've already determined to be so valuable.

You choose by assessing why the behavior, good or bad, is occurring. Everyone knows Mrs. Barker is great at making the students laugh, but how exactly does Mrs. Barker do this? Why are there so many disruptions in Mr. Polerecky's class?

Sadly, workplace performance has historically lacked an assessment of why behavior occurs (Austin et al., 1999). Without this knowledge, you're coaching blind, relying on a time-wasting trial-and-error approach. Let's illustrate the importance of this type of assessment through the following two examples.

Example 1: The Trial-and-Error Approach

In the halls of Westbrook Middle School, Principal Anderson notices Ms. Bennett's students are consistently underperforming. Puzzled and somewhat frustrated, she approaches the situation with the best of intentions, but her toolbox is painfully devoid of diagnostic precision. "We need to get those test scores up," she implores as she lays out a barrage of professional development workshops and general strategies that have seemingly worked elsewhere.

Ms. Bennett, already drowning in the sea of curricular demands and student behavioral issues, tries to implement the broad-strokes advice. She attends every workshop, scribbles notes furiously, and tweaks her lesson plans. Weeks turn to months, yet the needle barely moves. The students' engagement is still as lackluster as their test results. Principal Anderson is left scratching her head and questioning the commitment of a teacher who seemed to be doing all the right things but to no avail.

Example 2: The Informed Intervention

Principal Anderson takes a deep breath and sets aside her initial assumptions about Ms. Bennett's performance. This time around, she reaches for the Performance Diagnostic Checklist, a new tool she has just been trained in by a behavior analyst. She hopes this tool will be a compass to navigate the murky waters of underperformance. Before she sets foot in the classroom again, she combs through the checklist and works to understand the *why* behind the *what*.

Is it a skills deficit or a motivational issue? Are the expectations clear? Does Ms. Bennett have the resources she needs? As Principal Anderson observes the classroom, has discussions with Ms. Bennett, and reflects on the information gleaned, patterns begin to emerge. It isn't a lack of effort on Ms. Bennett's part; it is a misalignment of resources and reinforcements.

Armed with this newfound insight, Principal Anderson initiates tailored Deliberate Coaching sessions focused on effective classroom management and evidence-based instructional strategies. She ensures Ms. Bennett has access to high-quality curricular materials and arranges for a mentor teacher to provide in-the-moment feedback. This targeted support ignites a transformation, not just in the teacher's methodology but also in her confidence and engagement.

The results are palpable. Students begin to participate more actively, test scores see a genuine uptick, and Ms. Bennett's classroom becomes a model of improvement. Principal Anderson realizes that by taking a behavior-analytic approach, she is able to move beyond symptomatic solutions to systemic, sustainable change. Through deliberate, data-informed coaching, they turn a narrative of struggle into one of shared success.

Thankfully, for Principal Anderson and many others, the science of behavior has made some advancements in this area over the years, most notably with the development and extended application of the Performance Diagnostic Checklist (Austin, 2000; Austin et al., 2005; Wilder et al., 2018). The Performance Diagnostic Checklist (Figure 12.1) is a short questionnaire that assesses what elements in the workplace are causing performance problems to occur. Well-researched versions of this assessment tool have been applied to human-services settings (Carr et al., 2013) and safety (Martinez-Onstott et al., 2016). The foundation of the various checklist versions spans four areas, based on the antecedent-behavior-consequence model that encompasses behavior change: antecedents and information, equipment and processes, knowledge and skills, and consequences.

- **Antecedents and information.** Can the teacher explain to you exactly what they're supposed to be doing? Are there aids and prompts available to help the teacher know when and how to do it?

  These are types of antecedents that help get the behavior going. Educators need to have clear goals and expectations if you want them to meet your standards.

- **Equipment and processes.** Does the teacher have reliable materials in the classroom? Are the classroom and school processes organized and efficient, or complex and confusing? Are there any obstacles in the classroom that keep the teacher from doing their job well?

  School leaders need to set teachers up for success. You can't criticize a teacher for underperforming readers in the classroom if there are no books to read!

- **Knowledge and skills.** Can the teacher demonstrate the skill you're requiring of them? Can they do it reliably in the classroom environment when it's required? There is a big difference between attending a professional development event and actually demonstrating a skill. The number and quality of interactions during training, in particular opportunities to participate and to receive feedback and reinforcement, are critical to learning (Bucklin et al., 2000; Johnson & Rubin, 2011).

Think about some of the training you've received. How much did you and your peers have to participate? Were you required to demonstrate mastery or fluency? If not, that training event most likely did little more than make you generally aware of the topics under discussion and didn't actually expand your day-to-day repertoire with new skills.

If the teacher can't demonstrate the skill when asked, then you don't have anything to coach. You need to shift your focus to training those skills, then coaching the teacher to use those skills in the classroom.

- **Consequences.** Once someone can demonstrate the skill—and the environment isn't restricting them from consistently performing it at a high level—you're ready to make good things happen when that behavior occurs. Is the teacher being monitored? How often does the teacher receive positive feedback? What about constructive feedback? How often is constructive feedback followed up with positive reinforcement?

Antecedents such as training, task clarification, and prompts help set teachers up for success, but consequences show them that what you're asking of them actually matters.

Implementing the Checklist in Real-Time Coaching

The first three areas address whether someone actually *can do* the skill, whereas the consequences area addresses whether someone is capable of doing the skill but *won't*. For "can't do" (i.e., skill deficit), the intervention of choice should be linked to antecedent strategies such as training. Unfortunately, too often teachers are told what to do but not actually taught through concise instruction, opportunities to demonstrate the skill, and consistent feedback.

To determine if a teacher has a skill deficit, ask them to describe the skill, then ask them to demonstrate it. Remember, the question isn't "Were you trained?" or "Is there a procedure on this?"—the question is "Have I seen them do it?" Whether a teacher can demonstrate the skill and the situations in which they can demonstrate it effectively can guide whether more training is needed, more coaching is needed, or you need to better set them up for success based on improvements to their environment and the systems in which they work. If they can demonstrate a skill but they are lacking in speed, then the issue is one of fluency and simply requires more opportunity to rehearse speed and accuracy.

To determine if a teacher has a skill deficit, ask them to describe the skill, then ask them to demonstrate it.

A teacher can be well trained on how to address problem behavior, know all the policies and procedures to support these actions, and demonstrate the proper thing to do, but still do something else when a student acts out during class. Why? Because what the teacher is doing works. That is, there are powerful consequences keeping these behaviors going.

For instance, after meeting with a multidisciplinary team to review behavior data of a student, it is determined that the child's off-task behavior is being maintained primarily by escape from work. Specifically, the child is avoiding math he is clearly capable of doing based on previous assessments. For almost the entire day, he is very well-behaved. However, during the math block, the student reliably begins to drift off task and engage in a variety of other behaviors including drawing pictures, asking to go to the bathroom, sharpening his pencil, or sometimes attempting to quietly chat with a peer.

When these behaviors initially happen, the teacher quietly prompts the student back to task; however, after a while, the teacher becomes annoyed with the recurring behaviors and begins reprimanding. Initially this works as the student returns to task; however, this strains the relationship between the student and the teacher as evidenced by a new behavior. Noncompliance. In other words, the student eventually stopped returning back to task based on the teacher's reprimand, and even shouted at the teacher, "Why are you always picking on me!?" This incident leads the teacher to call the student's parents and request a meeting with the team.

After careful consideration, the team puts a plan in place that requires the teacher to break math assignments down into smaller chunks (e.g., providing half a page of math problems instead of a full page) and allows the student a brief break. In addition, the student is given permission to quietly draw pictures contingent upon completing a prespecified amount of work that will be progressively increased over time. The team thanks the teacher for calling the meeting and the coach reminds her, "Don't forget, these plans can take some time to work. Give it a chance."

So the teacher sets off with good intentions. She meets with the student to discuss the plan as it pertains to her role and his. Per the plan, she provides him with encouragement just prior to the math block as a reminder of the contingency (e.g., "Don't forget, first do this work, then you can draw for 3 minutes. I know you can do it!"). She then provides him with chunked work that he is to complete and return to her.

For the first week, the plan works beautifully. However, problem behaviors begin to rear their head again, which prompts the teacher to reprimand. As before, her reprimands initially work, but then quickly lose their effect as the student returns to being noncompliant. After a quick observation using the plan as a checklist, the coach realizes the teacher has omitted one pinpointed behavior in the plan—she stopped providing the student chunked work because it required increased effort. This led her to return to her primary behavior management tool that reliably worked for other students with infrequent misbehavior.

It wasn't because the teacher was a bad teacher. It's because this behavior management tool reliably worked in the past. In other words, good things happened as a result. Students stopped misbehaving and the teacher escaped situations that interrupted her instruction.

And ironically, chunking the work was actually far less effort than managing the misbehavior, far less disrupting to the classroom, and far less frustrating to the teacher! When the teacher breaks protocol, it's the job of the coach to help the teacher recognize that better things happen when they follow protocol. In other words, the intervention of choice for a motivation deficit should help them get in touch with positive outcomes as a result of correct performance. For example:

- Providing the teacher with more frequent and reinforcing feedback: "I notice you are following the plan by doing X; you are doing great!"

- Helping the teacher see the positive impact of their behavior: "As a result of doing X, have you noticed the student is completing more work and disrupting less?"

- Providing corrective feedback that is received by the teacher as helpful: "I notice you chunked the student's work the first week, but you aren't doing it now. I understand this might seem like more effort, but remember, it's costing you far less frustration and time as compared with managing the misbehavior. You can do it! Would you like help chunking some of the material this week?"

In the scenario above, if the student had been acting out because he has a skill deficit related to math, the entire meeting and intervention would have been a waste of time. Moreover, the student would have likely continued acting out, perhaps with increased frequency and intensity; the teacher would have likely gotten more frustrated and disillusioned; and the coaching relationship may have been negatively impacted as the solution supported by the coach did not work. Figure 12.1 shows a version of the Performance Diagnostic Checklist adapted for education (Gavoni & Costa, 2023). If the folks you're coaching aren't performing well or achieving desired outcomes, this checklist will help you intervene effectively by first allowing you to determine the root cause of a performance issue—prior to determining appropriate coaching interventions.

Performance Diagnostic Checklist—Education

Teacher Name: _____ Interviewer: _____ Date: _____

Briefly describe performance concern:_____

_____

_____

_____

Antecedents and Information	YES	NO
1. Is there a written description stating the clear expectation of the educator regarding a particular instructional/behavioral strategy?		
2. Has the educator received adequate instruction about what to do (e.g., instructions such as "I want you to do this and this before we leave today")?		
3. Has the educator received formal training on this instructional/ behavioral strategy? If yes, check all applicable training methods. ☐ Instructions ☐ Demonstration ☐ Rehearsal		
4. Is there a task aid visible *while* completing the instructional/ behavioral strategy in question (e.g., reminders to prompt the strategy correctly at the right time/duration)?		
5. Can the educator state the purpose of the instructional/ behavioral strategy?		
6. Is the educator verbally, textually, or electronically reminded to use the instructional/behavioral strategy? If yes, how often? ☐ Hourly ☐ Daily ☐ Weekly ☐ Monthly By whom? Check all that apply: ☐ Peer ☐ Coach ☐ Administrator ☐ Other		
7. Are there frequently updated, challenging, and attainable goals the educator is comfortable with about the instructional/ behavioral strategy?		
8. Is the educator "aware" of the school's mission?		

Equipment and Processes	YES	NO
9. If equipment is required, is it available and in good working order (e.g., computer, A/V, mic, etc.)?		
10. Are the equipment and environment optimally arranged in a physical sense (e.g., the arrangement of the students' desks)?		
11. Are larger processes performing well despite incorrect instructional/ behavioral strategies (e.g., routines and procedures)?		
12. Are these processes written out and arranged logically?		
13. Can the educator implement the instructional/behavioral strategy without any obstacles (e.g., interruption by the intercom)?		

Knowledge and Skills-Training	YES	NO
14. Can the educator tell you what they are supposed to be doing and how to do it?		
15. Can the educator physically/verbally precisely demonstrate the instructional/behavioral strategy?		
16. If the instructional/behavioral strategy needs to be completed quickly, can the educator perform it at the appropriate speed?		

Motivation	YES	NO
17. Are educators motivated based on the outcomes following the completion of the task?		
18. Do educators see the positive effects of implementing the instructional/behavioral strategy (e.g., increased student engagement, increased assessment data, decreased misbehavior)?		
19. Do administrators monitor the educator's behavior related to the task? If yes, how often? ☐ Hourly ☐ Daily ☐ Weekly ☐ Monthly		
20. Does the educator receive feedback about their performance? If yes, By whom? _____ and How often? ☐ Hourly ☐ Daily ☐ Weekly ☐ Monthly How long of a delay between observing the instructional/behavioral strategy and delivering feedback? _____ Check all that apply: Feedback Focus: ☐ Positive ☐ Constructive Feedback Type: ☐ Written ☐ Verbal ☐ Graphed ☐ Other		
21. Is the instructional/behavioral strategy easy to implement?		
22. Do other instructional/behavioral strategies appear to take precedence over the targeted strategy?		

Figure 12.1. *The Performance Diagnostic Checklist-Education offers a road map to quickly assess whether people are set up for success, if there are barriers hindering desired performance, and whether there are systems of ongoing feedback and reinforcement in place. The results displayed in the checklist guides your next steps in selecting a solution. Adapted from Gavoni and Costa (2023).*

Bridging the Gap: From Evaluation to Empowerment in Teacher Performance

Recognizing the potential for misalignment between student behavior and teacher interventions illuminates a profound truth in educational practice: Merely understanding the symptoms is not synonymous with comprehending the underlying issues. This insight lays the groundwork for a significant paradigm shift—from traditional, retrospective performance evaluations to a more enlightened, proactive coaching methodology. The transition from the former to the latter is marked not by the passage of semiannual reviews but by the day-to-day nurturing of teacher competencies. We stand at the cusp of this transformative journey, understanding that the tapestry of a teacher's performance is woven in real time and is best understood—and indeed influenced—through consistent, behavior-focused coaching that aligns with the principles of Organizational Behavior Management.

The key is not to await the semiannual revelation of performance evaluations but to engage in an ongoing dialogue—a dynamic interplay of action, feedback, and reinforcement. It's this relentless pursuit of improvement, with leaders as coaches who serve as the guiding North Star, that ensures performance evaluations are not dreaded reckonings but affirmations of a trajectory already well known and traveled by the teacher. In this new dawn, where coaching is not an event but a culture, we see the incremental progression toward excellence with every interaction, every piece of feedback, every small victory in the classroom. Now, let us delve into the fabric of this approach, exploring how the critical essentials of coaching unfold within the educational arena to elevate teaching and learning to their highest potential.

In education, performance evaluations tend to occur semiannually. As a result, they have little impact on performance improvement and really serve as a documented summary of what the teacher should already know. Because of the time involved and the negative impact on staff morale, many believe the process actually impedes rather than helps performance.

Aubrey Daniels, the godfather of Organizational Behavior Management, suggests that the most satisfied teachers are those who know at the end of the day how well they performed, so there are never any surprises. In other words, when it comes time for their formal evaluation, teachers and staff should already have a good idea of how they're performing.

To help teachers reach their maximum potential, coaches should always remember these five essentials:

1. **Align outcomes with pinpointed behaviors.** Teachers should know exactly what result they are working toward, and exactly what they should be doing to achieve that result.

2. **Track performance.** Teachers should be provided with measurement that allows them to gauge how well they're performing. For example, on the academic side, there are a variety of measures, including formative assessments, standardized assessments, and daily behavior data. At the school level, leading indicators, including student attendance, staff attendance, benchmark data, and discipline data, can serve as measures of schoolwide progress.

3. **Don't forget the "why."** Figure out why teachers are doing what they're doing, both good and bad. They're doing it for a reason, and knowing what that reason is will help guide your coaching.

4. **Feedback.** Measurement isn't any good if the teachers don't know about it. In professional sports, a variety of data is collected on individual and team performance. Imagine a professional football team having no information about the score, where they are on the field, what down it is, or how much time is left in the quarter. It won't know which way is up! Measurement toward goals should be shared with teachers weekly, or no later than monthly. Feedback on their behavior should be provided as often as possible. The most successful supervisors and managers provide feedback as immediately as they can. Feedback should be behaviorally specific and provided at a 4:1 ratio. That is, four positive statements recognizing appropriate behavior or incremental improvements to one constructive statement providing information that helps the performer improve. Supervisors who attempt to correct behavior more than they attempt to recognize and reinforce it will find themselves surrounded by unhappy teachers.

5. **Reinforcement.** Celebrate success as teachers reach established subgoals and goals weekly and/or monthly; reinforce behaviors daily that lead toward those goals.

You might be thinking, "How can I get measurement and feedback to the group of teachers I manage or supervise?" Well, each of you will have your own approach. Some coaches have strong habits linked to the use of measurement and feedback. Still others are working on their habits. One simple strategy that's gaining momentum is called the Big Five Performance Management system (Ferguson, 2014). Here's how it works. Each month, teachers submit a monthly email message to their coach describing two things:

- their five most significant accomplishments from last month in a one- or two-sentence statement
- their five highest priorities for the current month

This process takes no more than 10 minutes per month, but it gives teachers a chance to tell their story and take credit for their contributions. As a result, as a manager, you will have the opportunity to affirm, guide, and even redirect their efforts toward goals that you've both agreed on and that are aligned in the appraisal system. While providing regular documentation of your attempts to support performance improvement, it also improves the quality and quantity of your coaching efforts. This strategy can be linked to your school's teacher appraisal and alignment tool, keeping your team members strategically focused on their highest priorities. You can require your reports to submit emails each month on the fifth calendar day to help with consistency and double down on the idea of the "Big Five." Figure 12.2 shows an example of what it might look like.

To: Isabel Jones   From: Sharonda Murphy

March Accomplishments

- Helped JM learn how to request a break

- Worked with the team to redevelop LM's goals

- Researched an engagement strategy (choral responding) and began using it

- Personally met with three families to update them on their child's progress

- Helped Mary, my colleague, learn how to develop a lesson plan

April Priorities

- Focus on asking rather than telling my students, to foster more critical thinking

- Seek out more feedback from leadership and colleagues on my performance

- Research another strategy related to engagement

- Schedule a meeting with three more families to update them on their child's progress

- Prepare a de-escalation training for the upcoming parent training day

---

Figure 12.2. The Big Five Performance management system offers a way for teachers to communicate with their coaches on their top accomplishments from the previous month and their priorities for the upcoming month. Adapted from Ferguson (2014).

As you can see, the process is simple. In this example, when April rolls around, the teacher simply needs to cut and paste the April priorities into the next email and reflect on their accomplishments as related to these goals, then set priorities for May. All the coach needs to do is provide a quick affirmation of accomplishments where warranted, then give guidance or redirection where necessary.

For example, "Thanks for a great March! Keep up the good work, especially with those parent meetings. You've really improved with your frequency of these. Don't forget, progress reports are due next week. It's our goal to get 100% of them in by April 8." In other cases, after you've affirmed any accomplishments, you might simply state, "I'm not sure I agree with the focus of your parent training for next month. Please come see me so we can discuss before you begin."

You can also include thought-provoking questions several times a year in your responses. For example, "In your Big Five this month, please answer the following question: If you ran this school differently, what would be one thing you'd do to improve our customer service?"

Instituting Effective Coaching Systems in Education

Moving into the next chapter, we confront the challenge of building coaching systems that work effectively within the educational sector. These systems must account for both the broad goals of the school and the detailed behaviors that occur in classrooms. The focus is on establishing consistent practices that address the organizational structure, the procedural framework, and individual performance metrics. We will investigate how to institutionalize these processes through coaching, ensuring they become ingrained within the school's culture. This approach aims to ensure that Deliberate Coaching isn't just an add-on but an integral part of the educational ecosystem, one that drives sustained improvements in both teaching quality and student outcomes.

Key Takeaways

- The Performance Diagnostic Checklist is a crucial tool developed to identify workplace elements affecting performance, with proven applications in human-services and safety settings.

- Good leaders set people up for success and support progress. Even the best baseball coach will fail if they don't give the players a baseball bat.

- Effective coaching must address four foundational areas: antecedents and information, equipment and processes, knowledge and skills, and consequences.

- Clear goals, expectations, and appropriate teaching aids (antecedents) are necessary for educators to initiate desired behaviors and meet performance standards.

- It is vital for leaders to ensure teachers have the necessary materials and are working within efficient processes; without the right tools and clear processes, even skilled teachers can struggle.

- Skill demonstration is a step beyond theoretical learning; coaching should transition to training if teachers cannot apply what they've learned in actual classroom settings.

- Positive and constructive feedback mechanisms are essential; they not only reinforce desired behaviors but also demonstrate to teachers that their efforts are recognized and valued.

- Misalignment between student behavior and teacher interventions necessitates a shift from retrospective evaluations to proactive, consistent coaching.

- Continuous coaching based on behavioral principles should replace semiannual performance reviews to nurture teacher competencies and foster real-time growth.

- Leaders must serve as constant coaches to ensure that evaluations become regular affirmations of well-known progress rather than infrequent and dreaded reckonings.

- Teachers should already have a clear understanding of their performance before formal evaluations; regular feedback and reinforcement eliminate surprises.

- To maximize teacher potential, coaches must align expected outcomes with specific behaviors, track performance diligently, understand the "why" behind actions, provide frequent and balanced feedback, and consistently celebrate incremental successes.

13

Systematic Coaching

A goal without a plan is just a wish.

—ANTOINE DE SAINT-EXUPÉRY, *THE LITTLE PRINCE*

The Deliberate Coaching approach is a targeted and systematic approach. It is planned and strategic but also dynamic, designed to build, adapt, and produce demonstrable results in key areas of the learning process and classroom. Our schools are filled with educational and support systems working together to produce educated and respectable young people. Because these systems are dynamic, our methods must be dynamic. Because these systems are demanding, our teacher-management approaches must be organized and strategic so that we don't lose sight of where we're going (outcomes and linked teacher behaviors) and how we need to get there (coaching).

A Consistent Path to Excellence

Skinner (1953) once described how a scientific approach to behavior "shapes behavior as a sculptor shapes a lump of clay" (p. 91). The coach as sculptor shapes the behavior of a particular teacher by consistently working to support that teacher toward an ultimate performance goal. Good teachers are not born. Educators do not begin their careers with an innate ability to get the most out of their students. These behaviors need to be built, molded over time.

Take something as seemingly simple as the affect someone displays when speaking to a child. The pitch, proximity to the child, words used, and so on all need to be developed, either naturally through contingencies in the environment (e.g., by the child's reaction) or by a coach. Someone's first attempt at interacting with a child might fail miserably. The child might not respond to the person, might walk away, or do something disruptive. So you work with the person, reinforcing small improvements until they can finally speak to the child in an appropriate manner. Shaping behavior must be deliberate, and it must be planned and consistent, or the previously reinforced behavior will go away (or, more formally, it will be extinguished).

The same goes for tasks that involve numerous interlinked behaviors. These behavior chains must be reinforced systematically and supported until you get to the final performance goal. Before a teacher can appropriately troubleshoot a struggling student's problem and assist, they must first get comfortable interacting with students, showing them that they care and have something of value to contribute. We sometimes say this is a period of time when a teacher "pairs" themselves with reinforcement, meaning that the student sees that good things happen when that teacher is around. Once this is done, the teacher needs to be able to identify struggling students and notice signs and opportunities for improvement. Then the teacher must be able to determine why the student is struggling, select the proper solution, and assist the student until progress is made.

Because behavior occurs through reinforcement, and because behavior stops if reinforcement stops, coaches must be mindful of how and when they're offering support to teachers. Educators don't have a lot of time during the day to spare, so in order for coaching to be consistent, it has to be brief. And to maximize performance and avoid extinction of new and critical behaviors, coaching must also be frequent.

Measure Twice, Cut Once

Systematic coaching is anchored in this principle, embodying a sense of direction and organization. Taking the time to thoughtfully assess and plan—understanding not only what to do, but how and why—promotes informed decision-making (Figure 13.1). This careful preparation is pivotal in the shaping process and ensures that each action as a coach is deliberate and effective, thereby optimizing the use of time and resources.

The Deliberate Coaching System

Chapter 1 introduced the concept of *systems* and how we all work in interlocking systems of people, processes, and resources collaborating to achieve a common goal. These systems have subsystems, all connected and working

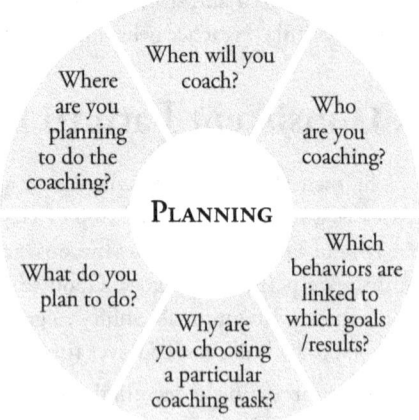

Figure 13.1. Instead of leaving the shaping of behavior to random chance encounters, the planning part of Deliberate Coaching offers a systematic road map to behavior change.

together like a fine-tuned machine. Each school is its own unique system that is part of a larger educational system in its district and region that also contains subsystems such as onboarding systems, student-testing systems, and training systems. Your onboarding system will get your teachers and staff acclimated to your school. Your training systems will produce improvements in skills needed on the job. But it is your Deliberate Coaching System where you'll put things into action, build habits, and improve culture. And the Deliberate Coaching System is built on the science of behavior and all the tools and resources that come with it.

THE FOUNDATION:
FIVE PHASES OF DELIBERATE COACHING

Deliberate Coaching is based on a proven, replicable process where performance is assessed, change initiatives are planned based on those assessments, behavior is set up for success with antecedents and supported through consequences, and progress is tracked and celebrated. The Five Phases of Deliberate Coaching are based on decades of research on the building blocks of behavior change; thus, they set the foundation of all leadership initiatives, including the Deliberate Coaching System.

Everything in this book—and frankly all behavior—changes the initiatives leaders engage in and fits within a framework of five key phases that make up this five-step process:

1. Assessment

When you conduct assessments in an educational setting, you gain clarity on the specific needs, strengths, and challenges within your school or its various subsystems. You are assessing the current link between critical pinpointed behaviors and valued pinpointed results throughout all applicable levels of the school system. This area also includes ensuring that all standards, goals, and change criteria are clear; that there is alignment between the job requirements and outcomes of each position in your school; and that there is alignment between school and district mission, vision, and values.

This phase involves organizational-level assessments, akin to analyzing the school's overall operational framework, including everything from curriculum design to student outcomes. It also includes process-level assessments that can identify gaps or inefficiencies in teaching methods or administrative processes. Lastly, it includes performer-level assessments tools such as the Performance Diagnostic Checklist to evaluate and enhance individual teacher or student performance.

2. Alignment

The alignment phase, as the second step following assessment, is pivotal in bridging the gap between the current situation and the desired outcomes in an educational setting (Figure 13.2). Based on the insights gained from the assessment phase, this step focuses on meticulously aligning behaviors with the desired results across all levels of the school system. It's about ensuring that every action, strategy, and objective is coherently linked to the overall goals and desired outcomes identified in the assessment. This means adjusting teaching methods, administrative processes, and even individual performance metrics to resonate with the school's mission, vision, and values.

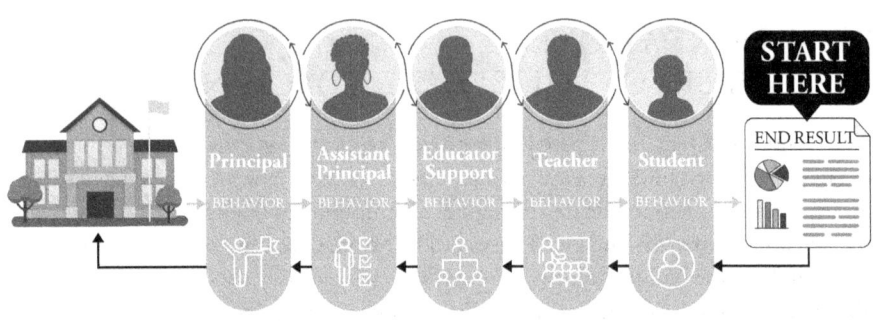

Figure 13.2. *This visual illustrates the strategic alignment of behavior across a school to produce intended results.*

The alignment phase transforms the "what is" identified in the assessment phase into a strategic "what should be," thereby setting a clear behavioral roadmap for achieving the school's broader educational objectives and enhancing its overall effectiveness. This process of realignment is critical for closing the gaps, optimizing performance, and driving meaningful progress toward the envisioned educational excellence.

3. Elevation

The elevation phase seamlessly follows the assessment and alignment phases and serves as a natural progression in enhancing educational performance. Building on the foundation laid in the assessment phase and the strategic realignments made in the alignment phase, the elevation phase focuses on empowering individuals and teams within the school system. This phase involves implementing key antecedent tools—such as comprehensive handbooks, targeted training programs, practical checklists, and supportive job aids—designed to address the specific needs and goals identified earlier.

By providing these resources, you create an environment that sets educators and students up for success and helps ensure that their behaviors are not only aligned with desired outcomes but also elevated to achieve peak performance and results. This step is critical in actualizing the potential uncovered and nurtured through the assessment and alignment phases and marks a pivotal step in realizing the school's vision of educational excellence.

4. Tracking

The tracking phase is the logical continuation of the assessment, alignment, and elevation phases, as it focuses on the meticulous observation and reinforcement of behavioral changes. This phase acknowledges that true behavior change doesn't happen overnight, nor does it wait for perfection. True behavior change requires consistent reinforcement of incremental improvements in performance. As a leader, you play a key role in keeping the momentum of change alive, ensuring that each small step toward the desired behavior is acknowledged and encouraged.

It is key to use a variety of tracking tools in this phase, including practical and adaptable resources like the Deliberate Coaching Self-Monitoring Tool (DCST) (see example in Figure 13.3 and template in the Appendix). This tool serves as a versatile instrument for both self-monitoring and direct observations, akin to a Swiss Army Knife for behavior tracking. By providing clear and tangible feedback, such tools aid in monitoring both immediate and long-term impacts of behavior changes on outcomes, ensuring precise reinforcement strategies. Incorporating self-monitoring techniques such as checklists and visual management systems enhances this process, allowing educators to maintain a dual focus on behavior change and its impact on educational outcomes. Additionally, the inclusion of leading indicators (e.g., behavioral changes) and lagging indicators (e.g., educational outcomes) as indices of progress ensures that the link between behavior change and its impact on educational outcomes is always visible and understandable, guiding the shaping plan effectively.

5. Communication

In the final phase, communication, the focus shifts to reinforcing behavior built through antecedents and celebrating results. As a leader, it's essential that you understand that the momentum of behavior change can fade away without consistent and supportive consequences. This phase is about accelerating performance and cementing positive habits by making sure that meeting the performance expectations established during the alignment phase leads to positive outcomes.

Effective communication tools play a vital role here. These tools include delivering constructive feedback, offering praise, and providing incentives. These tools are absolutely necessary for ensuring that individuals are not only aware of when they excel but also receive the necessary support and guidance when facing challenges. By consistently applying these consequence mechanisms, you reinforce the desired behaviors and outcomes and create an environment where positive change is recognized, celebrated, and encouraged. This phase completes the cycle and binds together all previous steps—from assessment and alignment to elevation and tracking—and ensures that the journey toward improved performance and results is acknowledged and reinforced through effective communication.

Deliberate Coaching Self-Monitoring Tool (DCST)

Coachee: Ms. Thomas   **Classroom makeup:** 20 Students   **Interval:** 15 minutes
Modality: Whole Group Instruction   **Coach:** Weatherly

Day/Date		Monday April 25			Tuesday April 26			Wednesday April 27			Thursday April 28			Friday April 29		
Whole Group Activities		1	2	3	1	2	3	1	2	3	1	2	3	1	2	3
Result: Increase student on-task behavior	**Scale** 5=all but 1 4=all but 2 or 3 3=Most 2=Half 1=<Half	5 4 3 2 1	5 4 3 2 1	5 4 3 2 1	5 4 3 2 1	5 4 3 2 1	5 4 3 2 1	5 4 3 2 1	5 4 3 2 1	5 4 3 2 1	5 4 3 2 1	5 4 3 2 1	5 4 3 2 1	5 4 3 2 1	5 4 3 2 1	5 4 3 2 1
Result: Reduce disruptions	5=10 4=8 3=6 2=4 1=<2	5 4 3 2 1	5 4 3 2 1	5 4 3 2 1	5 4 3 2 1	5 4 3 2 1	5 4 3 2 1	5 4 3 2 1	5 4 3 2 1	5 4 3 2 1	5 4 3 2 1	5 4 3 2 1	5 4 3 2 1	5 4 3 2 1	5 4 3 2 1	5 4 3 2 1
Behavior: Increase opportunities	5=6 pm 4=5 pm 3=3–4 pm 2=1–2 pm 1=0 pm	5 4 3 2 1	5 4 3 2 1	5 4 3 2 1	5 4 3 2 1	5 4 3 2 1	5 4 3 2 1	5 4 3 2 1	5 4 3 2 1	5 4 3 2 1	5 4 3 2 1	5 4 3 2 1	5 4 3 2 1	5 4 3 2 1	5 4 3 2 1	5 4 3 2 1

INCREASE ON-TASK

Defined: Student looking at teacher, assignment; interacting with peers according to activity expectations

Baseline: (3) Most (67% on task); Goal = (5) All but 1 (90% on task); Subgoal (4) Most (80% on task)

DECREASE DISRUPTIONS

Defined: Teacher temporarily stops instruction as the result of behavior during whole group activity

Baseline: (5) 8 disruptions; Goal = (1) 2 or less; Subgoal = (2) 4 or less

INCREASE OPPORTUNITIES TO RESPOND

Defined: Teacher asks academic questions seeking student responses during whole group activity

Baseline: (2) 1–2 pm; Goal = (5) 5 pm; Subgoal = 3–4 pm

Figure 13.3. Adapted from Iovannone et al. (2014), the DCST allows for a quick and easy way to track both immediate and long-term impact of behavior change.

THE DELIBERATE COACHING BRIEF: LEVERAGING PRECISION AND FREQUENCY FOR EFFECTIVE EMPLOYEE DEVELOPMENT

In a high-performance environment, time is of the essence, and the same holds true for coaching. At the intersection of behavior analysis and leadership lies our approach to employee coaching, an approach we refer to as "Deliberate Coaching Briefs" (see Figure 13.4). This is where you get to showcase your foundational understanding of behavior, and it is the avenue to making change. The term "briefs" is chosen deliberately; it encapsulates our philosophy that coaching interactions ought to be precise, purposeful, and systematic, and should focus on brief, impactful dialogues that drive meaningful behavioral change, not on elongated conversations.

Plan the Work and Work the Plan

Try to be quick in your interactions so as not to waste time. This will also allow you to increase the frequency of your coaching interactions. Everything you do as a coach funnels down to the coaching interaction. Your observations help you find your pinpoints: What is important to the classroom? What is important to the teacher? What are the behaviors that are working? What are the behaviors that are not working? You need this information to know where performance is and where it needs to be. You need it to guide your feedback statements, ensuring precision.

Your assessment tells you what you're supposed to do during your coaching interaction. Do you need to ask questions to find the reason for a performance deficit? Do you need to model to help build a skill? Do you need to provide constructive feedback to get a new behavior going? Do you need to provide any follow-up reinforcement? Everything in your assessment leads to your coaching interaction.

The Mechanics: How Does It Work?

In each Coaching Brief, the aim is to engage employees in a focused dialogue that directs their attention to key behaviors and the ensuing outcomes. Rooted in the science of human behavior, these interactions are tailored to promote self-observation and self-management. The coach's role isn't to provide long-winded lectures but to ask incisive questions based on the 8 Ws that catalyze insights into the employee's behavior and its impact. These questions can range from inquiries about task performance to deeper considerations about the alignment of individual actions with broader coaching goals. The 8 Ws are discussed in-depth in Chapter 10 and are the foundation to guide your coach planning, interactions, and quick tracking.

DELIBERATE COACHING BRIEF

TIMING · BRIEF · FREQUENT
CONSISTENT · CONTENT

Discuss outcome you're working toward
What goal are you trying to accomplish?

Discuss improvements
How did it work? What did you see?
Provide feedback (positive or constructive).

Discuss behavior the teacher engaged in
What did you do to change their behavior,
and how does this relate to the goal?

Discuss why it worked or why it didn't
Pinpoint behaviors responsible for the impact.
Troubleshoot solutions if performance
problem or outcome isn't being met.

Discuss next steps
What will you do next?
What do you need from me as the coach?

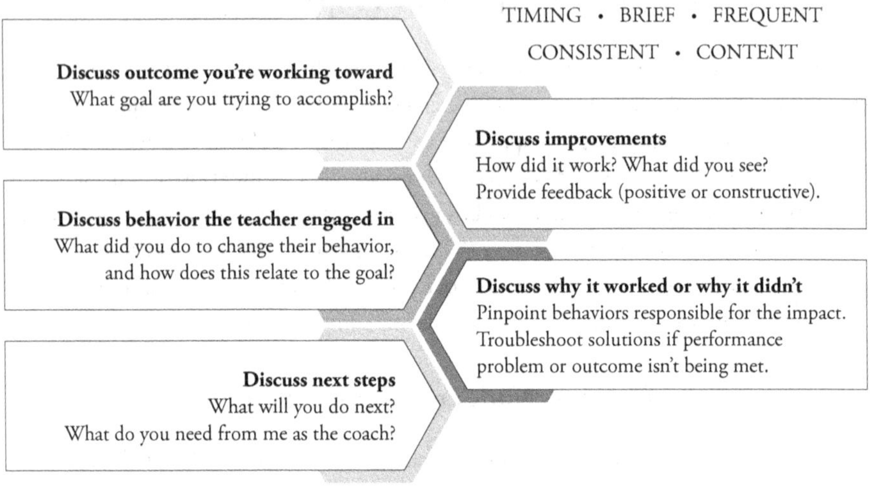

Figure 13.4. The Deliberate Coaching Brief is the key to shaping behavior through quick, concise, and frequent coaching interactions.

Why Brief?

So, why put a premium on brevity? The rationale is twofold. First, brief interactions allow coaches to spread their influence more broadly across the school and to reach more employees in a given period. Second, shorter, more frequent interactions promote a continuous feedback loop, which from a behavioral perspective, is far more likely to facilitate sustained behavioral change than would infrequent, lengthy sessions. The focus shifts from the duration of the interaction to its frequency. This allows for consistent reinforcement and offers room for immediate course corrections.

The Benefits: Quality Over Quantity

By centering on purposeful questions and acute observations, each brief coaching session becomes a crucible for self-awareness and behavioral refinement. It is not merely a time-saving exercise but a strategic approach to maximize the utility of each interaction. Employees are not just recipients of feedback but active participants in a process that heightens their observational skills, enhances their performance, and aligns them more closely with organizational goals.

The Bottom Line

Coaching Briefs are designed to be more than just quick check-ins; they are concentrated capsules of impactful coaching, rooted in the principles of behavior analysis. These sessions transcend the limitations of time, proving that impactful coaching is not about how long you talk, but how well you leverage each moment to instigate positive, lasting change. This methodology underscores our unwavering belief in the power of behavioral science to drive human performance to new heights, one brief but potent interaction at a time.

THE DELIBERATE COACHING MEETING: ACCOUNTABILITY AND SUPPORT

After establishing a robust framework through the phases of assessment, alignment, elevation, tracking, and communication, the Deliberate Coaching Meeting becomes a cornerstone for both accountability and support. These meetings are integral to the continuous development and refinement of coaching practices as well as accelerate performance improvement within and across classrooms.

In these meetings, coaches are expected to bring precision and focus to their interactions, moving beyond ad hoc approaches to a more structured and data-driven methodology. By consistently providing Coaching Briefs to their direct reports and employing tools such as the 8 Ws form, coaches ensure that their efforts are both targeted and measurable. These gatherings offer a platform for sharing experiences, discussing challenges, and celebrating successes in coaching. They serve as a conduit for "coaching the coaches" and reinforce the importance of adhering to effective coaching behaviors and techniques.

The goal is not just to fulfill an obligation but to find intrinsic value in these interactions and to discover the natural reinforcers that make these meetings a fertile ground for growth, learning, and professional development. Hence, Deliberate Coaching Meetings provide an opportunity for coaches to engage, evolve, and exemplify the very essence of impactful coaching.

Steps for the Deliberate Coaching Meeting

Before we dive into the nuts and bolts of what a Deliberate Coaching Meeting should entail, let's anchor ourselves in its undeniable significance. The essence here is to create a cyclical learning environment where coaches not only share but also absorb as they strategize real-world solutions based on hard metrics and peer insights. Done correctly, simply by following the 8 Ws form to plan your coaching, track your coaching, and as an aid to report out during these meetings, the time spent with leaders discussing good examples of leadership becomes a powerful incubator for best practices and innovative solutions, all while reinforcing a Deliberate Coaching culture that's precise, purposeful, and systematic.

Figure 13.5 unpacks the meeting process and provides you the "why" for each step to help you better understand how each connects to the bigger picture. See the Appendix for a step-by-step Deliberate Coaching Meeting Protocol.

INITIALIZATION (5 minutes)

- **Why:** To set the tone and clarify the objectives of the meeting.
- **Tasks:**
  - **Opening Remarks:** Brief overview of why this meeting is critical for skill development and sharing best practices.
  - **Agenda Overview:** Briefly outline the meeting structure to set expectations and ensure all participants are on the same page.

RANDOM COACH SELECTION (2 minutes)

- **Why:** To ensure diverse perspectives and equitable opportunities for sharing.
- **Task:**
  - **Selection Method:** Utilize a random selection process to identify the coach who will present.

REPORT-OUT SEQUENCE (10 minutes per coach)

- **Why:** To facilitate focused, constructive conversations and skill sharing.
- **Guiding Document:** Use the pre-filled Deliberate Coaching Summary Sheet (in the Appendix) to guide the conversation to ensure consistency and alignment with objectives.
- **Task:**
  - **Establish Ground Rules**—concise, precise, specific:
    This approach ensures clarity and depth and aids effective communication and comprehension.
  - **Structured Sharing (leader to peers/facilitator):**

    #### OBJECTIVE AND METRICS (2 minutes)

    - **Why:** To link real-world outcomes with actionable steps.
    - **Task:**
      - Present your target result and explain its significance for stakeholders.
      - Discuss the metrics you used.

    #### EXECUTION STRATEGY (2 minutes)

    - **Why:** To share the "how" and to offer actionable insights for peers.
    - **Task:**
      - Discuss what steps you took to meet your goals.

RESULTS ACHIEVED (1 minute)

- **Why:** To validate strategies and offer learning opportunities.
- **Task:**
  - Describe the outcomes, both successes and failures, and provide a complete picture of your efforts.

FUTURE COURSE OF ACTION (1 minute)

- **Why:** To lay out next steps for sustained progress.
- **Task:**
  - Discuss what you intend to do next based on the outcomes you achieved.

PEER COACHING AND FEEDBACK
(to the leader doing the structured sharing; 5 minutes)

- **Why:** To encourage mutual growth and improvement through community knowledge.
- Consider the following COACH acronym to guide how peers respond to a given leader's report-out:
  - **Celebrate and Capture (C):** Peers acknowledge and celebrate the achievements and successes shared by the leader. This step ensures that positive efforts are recognized and appreciated, fostering a supportive environment.
  - **Observe (O):** Peers attentively observe and listen to the leader's presentation, focusing on both verbal and non-verbal cues. Active listening enables peers to gain a comprehensive understanding of the leader's experiences and challenges.
  - **Ask (A):** Peers ask clarifying questions to deepen their understanding of the leader's strategies, outcomes, and future plans. These questions encourage reflection and promote a deeper exploration of the presented material.
  - **Constructive Feedback (C):** Peers offer constructive and actionable suggestions for improvement, if warranted. This feedback aims to help the leader refine their approaches, overcome obstacles, and achieve their goals more effectively.
  - **Highlight (H):** Peers highlight key insights, learnings, or points of resonance from the leader's presentation. This step ensures that valuable lessons are captured and shared among the group for future application and continuous improvement.

CLOSING REMARKS (5 minutes)

- **Why:** To reinforce key takeaways and set the stage for continuous learning.
- **Summary of Learnings:** Recap the session and emphasize the actionable insights gathered.
- **Preview of Next Meeting:** Offer a brief look ahead to foster ongoing engagement and preparation.

Figure 13.5. The agenda for your Deliberate Coaching Meeting offers key objectives and targets to ensure a fluid and productive engagement.

By methodically structuring the meeting this way, we shift away from the "meeting for meeting's sake" trap and toward a model of communal growth and deliberate improvement. Sound like a plan?

Deliberate Coaching Meeting in Action

Having outlined the key steps of the Deliberate Coaching Meeting, it's now time to bring these concepts to life through a practical vignette. This illustrative scenario will provide a glimpse into how the principles and practices of Deliberate Coaching are applied in a real-world setting. It serves as a tangible example, demonstrating the dynamics, conversations, and strategies that embody effective coaching in action. Through this vignette, we will see how the theoretical framework of Deliberate Coaching materializes into practical application, offering insights into the nuances of coaching interactions, the implementation of structured methodologies, and the real-time problem-solving that occurs in these meetings.

LAKESIDE ELEMENTARY

Picture this: It's another Tuesday afternoon at Lakeside Elementary School, and the usual conference room has transformed into a hub of targeted professional growth. Filling the seats are instructional coaches, team leaders, and the principal, each armed with a Deliberate Coaching Summary sheet. Zara, a coach known for delivering results, sets the tone, reminding everyone, "We're diving deep, but we're doing it efficiently. Concise, precise, and specific—that's our mantra today."

First up, randomly selected, is Mark. "I focused on elevating on-task behavior among the sixth-graders in Mr. Jackson's class. I measured success by the rate of completed assignments," Mark says, glancing at his summary sheet for reference. "We all get why this is vital, right? More learning for the students, less stress for the teachers."

Mark then outlines his intervention strategy with Mr. Jackson, focusing on the behavior-specific cues and the rewards that followed. "Within a week, completed assignments increased by 20%," he declares. Nods of affirmation circulate the room. He finishes off with his plans for sustaining the improvement.

"Alright, time to coach the coach," announces Zara. "What struck a chord with you from Mark's sharing?"

Jennifer is quick to jump in, "Your approach to behavioral cueing really caught my attention. Do you think this could be modified for kindergarten behavior issues?"

Oscar, the principal, chimes in, "I like the immediate metrics you got from this approach. Were you surprised by the speed of the improvement?"

Mark considers the questions and responds, offering deeper insights into his methods and their adaptability.

Zara then asks, "Any pinpointed suggestions for Mark?"

Oscar suggests, "Maybe adding a parental element to the reward system could further reinforce the behavioral changes."

As Mark jots down the suggestions, a wave of applause ensues. Zara wastes no time. "Let's keep the ball rolling," she declares.

It's evident to everyone present that this isn't just another meeting—it's a finely tuned coaching exercise, brimming with real-world value. Each coach leaves not just heard but enlightened, ready to apply their newly gained knowledge. This is Deliberate Coaching in action: concise yet comprehensive, individual yet collaborative. It results in a collective efficacy that's bound to radiate across the classrooms. Doesn't this beat your typical staff meeting?

From a Behavioral Perspective

From a behavioral perspective, this Deliberate Coaching Meeting process isn't merely a meeting; It's a vital performance management intervention. What makes this so pivotal for success?

Let's start with the principle that organizations, at their core, are comprised of individual behaviors strategically aligned to meet specific objectives. The Deliberate Coaching Meeting is designed to hone these individual behaviors. The use of the 8 Ws form provides a mechanism for clarifying performance metrics, capturing behavioral data, and defining success—core tenets of behavior analysis.

By structuring the meeting to share and discuss successful experiences, we are capitalizing on the power of social reinforcement and vicarious learning. Colleagues learn not just from their own experiences but from the validated experiences of others. This spreads effective practices faster than individual trial and error ever could, maximizing the utility of both human and organizational resources.

The time-structured reporting, centered around results and action steps, aligns perfectly with the behavior-analytic focus on actionable, results-oriented behaviors. It's not just about what you think or know; it's about what you do and how that action contributes to organizational goals. The steps for continued support ensure that the coaching process is cyclical, not linear. This aligns with the behavioral principle of continuous performance management, driving perpetual growth and adaptation.

Finally, the "coaching the coach" segment provides an immediate feedback loop—another critical behavioral component. This not only empowers coaches with real-time performance data but also encourages accountability and ongoing self-assessment. This reinforces the desired behaviors at an individual level while collectively moving the organization toward its objectives.

And let's not forget, behavior science thrives on data. The entire Deliberate Coaching process serves as both an input and output of actionable data, which is gold for any organization looking to base its decisions on empirical evidence rather than gut feeling or tradition.

To sum it up, you can't afford not to engage in a systematic coaching approach like this. The risks? Minimal. The rewards? Potentially monumental.

One System to Unite Them All!

You now have the scientific foundation, the tools, the processes, and the phases, all of which are under an umbrella of your Deliberate Coaching System (Figure 13.6). It all fits together! One system that encompasses the critical components necessary for any behavior change initiative, whether in the workplace, home, or anywhere else.

If you stick to the basics within the Deliberate Coaching System, not only will this align with your other leadership systems such as your training and evaluation systems, but it will help you use your tools, processes, and other systems of leadership and staff development more efficiently and effectively. Remember, trial and error is wasteful, and time wasted during a school year is valuable. We can't get that time back.

Leaders, as with all staff, need clear expectations, training support, and the appropriate time and resources to do their jobs at a high level. Leaders need to be set up for success in how to lead. Accountability is critical to ensure there is alignment on expectations and goals, but people do what they do because it's working for them. Once they are set up for success, their progress needs to be encouraged; they need to see the value in meeting your school's performance expectations.

Remember, trial and error is wasteful, and time wasted during a school year is valuable. We can't get that time back.

All systems of support, reinforcement, tracking and evaluation, accountability, peer-to-peer coaching, and so on are found under the umbrella of your unified leadership system—your Deliberate Coaching System! You might not be in full control of all facets of the various systems in your workplace and communities; however, you *can* influence your systems and you *can* control what you do. This is your starting point, and *your* Deliberate Coaching System can help you kick things off!

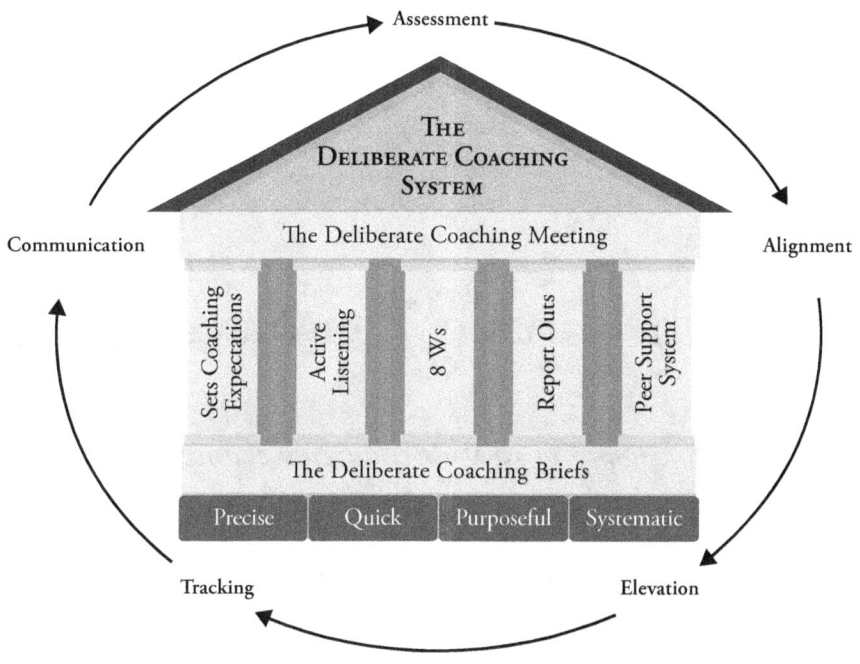

Figure 13.6. *The Deliberate Coaching System includes everything you need to drive transformative culture change, setting people up for success, using Deliberate Coaching Briefs to shape behavior, and building ongoing systems of support through the Deliberate Coaching Meetings, all based on fundamental principles of behavior analysis.*

Key Takeaways

- Systematic coaching embodies a sense of direction and organization, and taking the time to thoughtfully assess and plan—understanding not only what to do, but how and why—promotes informed decision-making.

- The foundation of Deliberate Coaching centers on five phases: assessment, alignment, elevation, tracking, and communication.

- Sustainable culture change occurs when leaders hold other leaders accountable for reinforcing and supporting behavior change; set those leaders up for success; and positively reinforce their progress, not just their impact. Deliberate Coaching Meetings offer an avenue of a quick system change that can offer this support and involves leaders learning from leaders.

14

Being the Coach You Need to Be

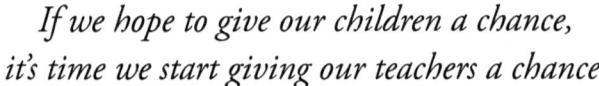

If we hope to give our children a chance,
it's time we start giving our teachers a chance.

—BARACK OBAMA, 2005, CENTER FOR AMERICAN PROGRESS

As you're considering how to bring together all you've learned, there must be a first step. Your first step may be to set initial goals. However, you may have an inner voice telling you, "I'm too busy." Still, you might try a few things out. Perhaps you'll show a few of these tools to your colleagues. But soon you may fall back into your old habits. If you find yourself there, remember this: People don't reach goals, no matter how small, when they fail to consistently engage in the behaviors that will lead them there.

If you value something, you must first learn to recognize the behaviors that will lead you toward your goals, as well as recognize the behaviors that will lead you away from them. Reaching goals and successfully applying Deliberate Coaching principles with an individual, grade group, or school requires you to begin by deliberately doing something more, less, or differently. When you do this, you can effectively help the teachers you are supporting do the same to achieve their own success.

Remember, all results are achieved by behavior. It's not magic. It's simply deliberate and consistent behavior. Lack of time is often not the root cause of failing to reach goals—it's typically what people *do* with their time that leads them to come up short. So, if you want to achieve goals and help those you are coaching be

> *People don't reach goals, no matter how small, when they fail to consistently engage in the behaviors that will lead them there.*

successful, get out of your mind (e.g., thinking "I'm too busy") and get into your behaviors, one step at a time.

When determining which behaviors to begin with, assess your current strengths and look for areas in need of improvement. There are things that you do as a coach that teachers like and respond to, and that produce desired results. You need to continue these things. There are also small improvements you can start making. For example, you might need to adjust how often you see the teacher, engaging in more frequent, quick interactions and fewer long, infrequent feedback sessions. You also might consider how specific you are when you speak about teacher performance to clarify the behaviors you wish to change.

We've offered a self-assessment tool (Figure 14.1) that you can use to help identify some strengths and opportunities in your individual coaching. The assessment is broken down into areas of coaching, providing feedback, advising, and development planning. Your score can help guide you on the good practices you need to integrate into your long-term coaching.

COACHING SELF-ASSESSMENT AND CHECKLIST

Perhaps you are new to the concept of coaching, or maybe you are a seasoned pro. Either way, use this sheet to periodically check yourself against core competencies that are essential to building a culture of coaching.

Point scale: 0 = never, 1 = only when requested by employees, 2 = inconsistently, 3 = at regular intervals (once or twice a year), 4 = as often as possible.

Please rate your skills on the following questions. Then, star two areas where you feel you have strength as a coach, and check two areas where you would like to improve your coaching.

Coaching (listening and drawing out)

Talk with employees about what they like most about work (not necessarily just their current job).	
Listen to employees' concerns about the progress and direction of their careers.	
Learn what is motivating for your employees.	
Step out of the "expert" role and truly listen, just to understand your employee's perspective.	

Reality testing (providing positive and corrective feedback)

Recognize and reward employees for their contributions, in the presence of others.	
Initiate discussions with your employees about how you see their future career.	
Explain formal and informal factors that determine success in the organization.	

Advising (offering organizational insight, information, and advice)

Talk clearly about changes in the organization that will require new learning.	
Offer suggestions for opportunities for new learning.	
Work to identify on-the-job opportunities for your employees (inside and outside your immediate area) that are in line with organizational goals and employee desires.	
Provide resources for employees who want information on developmental opportunities.	

Development planning (guiding employee goal setting, following up)

Review and discuss an employee's written professional development plan and goals.	
Discuss expectations (yours and theirs) regarding professional development planning.	
Check in with employees about their progress on development goals/initiatives.	
Provide physical support (funding, time off, personal assistance) for employees to pursue more learning.	
Provide motivational support for employees to pursue more learning.	

Figure 14.1. *Self-assessment offers a way to understand your strengths, your progress, and opportunities for improvement. The Coaching Self-Assessment and Checklist is a quick guide to check yourself against core competencies.*

Social Validity in Deliberate Coaching

As we begin to wrap up Deliberate Coaching, it's important to recognize that the success of coaching extends beyond just achieving tangible results. Imagine a scenario where a school leader, proud of the outcomes they've facilitated, overlooks a vital aspect of their approach: the perceptions and experiences of the person being coached.

While a specific result may have been achieved, if the methods employed or the journey to that result did not resonate positively with the teacher, the long-term effectiveness and sustainability of the coaching are called into question. This highlights the need to determine whether the coaching process is not only effective in its outcomes but also aligns with the values, preferences, and perceptions of those being coached. It's not just about what *you* want and need. It's also about what *they* want and need. As such, one of the guiding principles behind Deliberate Coaching is *social validity*. In the world of behavior analysis, social validity is defined by the following three components (Wolf, 1978):

- social significance of the goals of treatment
- social appropriateness of the treatment procedure
- social importance of the effects of treatment

You might be thinking, "What the heck does *treatment* have to do with coaching? Treatment to most folks usually means a doctor treating a patient to help them get better." Well, let's take a deeper albeit quick look at the term. Dictionary.com (n.d.) defines treatment, in part, as:

1. An act or manner of treating.
2. Action or behavior toward a person, animal, etc.
3. Management in the application of medicines, surgery, etc.

If we take a few liberties and substitute "coaching" in the definition, it looks like this when applied to education:

1. An act or manner of coaching a teacher.
2. A coaching action or behavior toward a teacher.
3. Management in the application of coaching.

It's easy to see how Deliberate Coaching might be considered a type of treatment, as it's an approach and application of strategies intended to help teachers perform better. Let's return to the concept of social validity. Carter (2010) defines it as, "The evaluation of the degree of acceptance for the immediate variables associated with a procedure or program designed to change behavior" (p. 2). Simply put in terms of Deliberate Coaching, it's a measure of a person's reaction to or perception of the coaching. In order to continue applying Deliberate Coaching, not only must the coaching be effective (i.e., have a positive impact on student achievement), it must be considered relevant to the teachers, be provided in a way that engages them, and produce outcomes that are valued by the stakeholders.

So if we return to the above three components as defined by Wolf (1978) but substitute the term "treatment" with "coaching," it meets the criteria we spoke of and looks like this:

- social significance of the goals (pinpoints) of coaching
- social appropriateness of the coaching procedure
- social importance of the effects of coaching

If your Deliberate Coaching approach isn't socially valid, you won't bring out the best in the teachers you serve. In fact, you might be lucky to simply get compliance, and you'll only get that when you're there providing direction. As soon as you leave the room, they'll likely go back to what they were doing before you stepped foot in their classroom. Because of this, it's important that you measure the reaction or perception of those you're coaching so you can continue to monitor and improve your Deliberate Coaching.

How can you determine the social validity of your coaching? Easily—ask those you coach! Here are some tools to help you along the way, based on *The Social Validity Manual: A Guide to Subjective Evaluation of Behavior Interventions* (Carter, 2010; Figures 14.2 and 14.3). You can provide these to individual teachers and use the aggregate data to determine trends related to how people perceive your coaching.

Rate your agreement/disagreement with each of the statements below for the following coaching pinpoints.						
COACHING PINPOINTS	**1**	**2**	**3**	**4**	**5**	**6**
This is the best coaching pinpoint that could be chosen.						
This coaching pinpoint focuses on the most important issues.						
This coaching pinpoint increases the opportunity to improve in other areas.						
This is a reasonable coaching pinpoint to accomplish.						
This coaching pinpoint will not have a negative impact.						
This coaching pinpoint will increase opportunities for reinforcement.						
This coaching pinpoint is needed more than most other pinpoints.						
1 = Strongly Disagree; 2 = Disagree; 3 = Slightly Disagree; 4 = Slightly Agree; 5 = Agree; 6 = Strongly Agree						

Figure 14.2. Determining social significance of the coaching pinpoints. Based on Carter (2010).

Rate your agreement/disagreement with each of the statements below for the following coaching goals.						
COACHING GOALS	1	2	3	4	5	6
I feel confident in my understanding of the coaching procedures.						
The coaching procedures are acceptable and ultimately linked to student outcomes.						
I am willing to follow through on the coach's recommendations.						
Given your needs and the needs of the students in my classroom, I find the coaching to be reasonable.						
The coaching I am receiving has been effective toward improving pinpointed skills.						
The coaching I am receiving is having a positive impact on pinpointed results.						
Following the coach's recommendation requires little effort from me.						
1 = Strongly Disagree; 2 = Disagree; 3 = Slightly Disagree; 4 = Slightly Agree; 5 = Agree; 6 = Strongly Agree						

Figure 14.3. Social appropriateness and importance of the coaching. Based on Carter (2010).

As you've learned, Deliberate Coaching requires purpose or intention. That said, always remember this: People won't judge you by your intentions, but rather by your impact. While you may possess strong knowledge and good intentions, your coaching needs to be received well by those you're coaching. Social validity measures help you determine if it is, and this will guide your approach.

Coaching the Coaches

It's important to note that the Deliberate Coaching process can and should cascade through all levels of the school and even the district (see Figure 14.4). If you are a senior leader in your district or school, you will have likely identified school outcomes that are important, aligned those school outcomes with classroom outcomes, and aligned those classroom outcomes with the behaviors that teachers need to engage in to achieve them. Hopefully, you've gone on to implement a Deliberate Coaching System where the leaders in the school who are responsible for teacher performance use

Cascading Deliberate Coaching

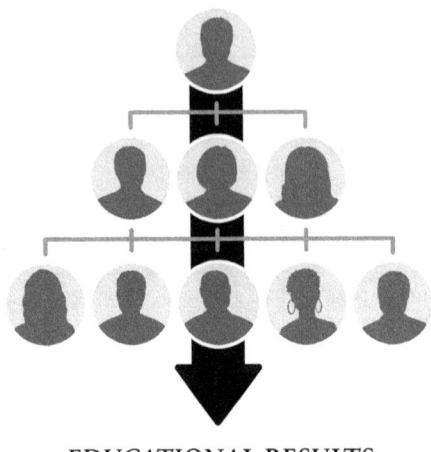

EDUCATIONAL RESULTS

Figure 14.4. Coaches also need coaching to ensure they're set up for success and receiving ongoing support. Deliberate Coaching is a cascading process across all levels of a given system.

tools such as the 8 Ws to frequently and consistently support each teacher's progress toward hitting their classroom goals.

But don't forget about supporting the leaders who support the teachers! Any new leadership tool, process, or overall system is going to take some getting used to. Even though we recommend starting small—by simply trying to tweak how leaders interact with teachers and then moving to frequency, consistency, quality, and the use of tools—this is still a new process. And new processes mean you're taking leaders out of their comfort zone, away from things that have occurred for a while because they've seen some sort of value in those things, and asking more of them. This adds a level of effort and uncertainty, both of which are going to compete with the new coaching behaviors you're looking for (those outlined in this book) and make the current behaviors you're trying to improve seem more appealing. In other words, change can be tough, so it's easy to stay with what you're currently doing.

Those responsible for the performance of the coaches need to recognize this and be the leader they need to be based on the needs of their coaches. Don't forget that those doing the reinforcing need reinforcement too. The behavioral principles cascade as well. While your Deliberate Coaching System should provide coaching the coach opportunities, it's important that this is not the *only* coaching that they will receive.

REVIEWING THE 8 WS

With that in mind, let's revisit the all-important 8 Ws that you should be using before, during, and after the Deliberate Coaching Meeting!

Who?

Who are you coaching? Perhaps you're a principal coaching a vice principal or a vice principal coaching a teacher. Or an executive director coaching a principal.

Why?

Why are you choosing a particular coaching task? You might be observing the vice principal providing feedback so you can see exactly what components she's giving during the feedback conversation and how the teacher is reacting to the feedback. Or your target coaching behaviors for that week could be assessing what's working and what's not working with the vice principal to find out why the coach might be struggling in a certain area.

Just as when you're coaching the teacher, this is important because you want to maximize the use of your time. You do this by ensuring that this is truly a skill-deficit problem and not a problem that should be addressed with prompting or consequences. Most likely there will only be a few people available to coach at your school and a lot of teachers who need coaching, so maximizing (and limiting) how much time you spend with them will be appreciated!

Which?

Which behaviors are linked to which goals and results? This book has identified a number of behaviors that are associated with successful coaches. A coach is indirectly responsible for the results of a classroom and directly responsible for the behaviors of the teacher. Which of these behaviors are your targets during your coaching for this particular week? What teacher outcomes are linked to this behavior?

Perhaps you're focused on the quality of the feedback statement provided from a vice principal to a teacher, or whether the vice principal follows up with the teacher as planned after previous constructive feedback. Their providing feedback might be important because the teacher's performance is linked to a particular student-achievement area. Keeping this alignment when you're coaching the one doing the coaching helps you and that coach understand what you're supporting and why this is important.

When?

When will you coach? Based on the aforementioned issue of time constraints, it's important to figure out the best time for your coaching. The sooner you can meet with the coach after the coach has an opportunity to engage in the behavior you're targeting, the better. Remember that delayed consequences are less effective, and by delaying you also make it more likely that the coach will forget important details about the interaction. If you're not directly observing the coach in action, then what you're left with is reinforcing their account of that interaction. You need reliable information.

Where?

Where are you coaching? If you can't observe and engage with the coach in the environment they're coaching in without sacrificing the integrity of their coaching and privacy, then arrange to have a private place where you can meet shortly after. You don't want to accidentally find yourself providing feedback and having a coaching discussion in public. This will backfire quickly, even if you're simply providing praise.

What?

This is the point in the 8 Ws where you're planning for what you'll be doing and also recording what you *actually* did and using it to report out during your Deliberate Coaching Meetings. Remember that this form is supposed to be used as an antecedent to guide your coaching, helping you hit the steps that are critical for success. For example, it might say that you're supposed to find the vice principal on Wednesday after her observation session with Mr. Clark and ask questions about how this session went in order to find opportunities for positive or constructive feedback. Once the coaching interaction is completed, you'll track what you actually ended up saying. This gives you a starting point for assessing how your coaching went and acts as a record in case *your* supervisor would like to discuss *your* coaching!

What Happened?

What was the impact? How do you know that the coaching has been successful? Part of tracking your process is ensuring that you know whether your coaching is working. This can be a potential reinforcer for you, and it's a way for you to show your supervisor the success you're having. It also gives you something extra to focus on if you need to discuss various areas of success with your vice principal or whatever coach you're supervising.

Your impact includes various results and outcomes, including the behavior of the coach you're coaching, the behavior of the teacher they're coaching, and any classroom changes you might be seeing. It also should include any notes you have on how the coach is taking your coaching. Do they seem to be asking you more questions? This is good. Do they seem to be approaching you more to share examples of their coaching? This is *very* good! How is their body language? Are they deflecting your feedback?

The direction of their behavior change will tell you whether you're having a reinforcing or punishing effect. These additional notes will help paint a picture of how much they're valuing your coaching conversations.

Why Don't … ?

"Why don't you follow up by prompting the teacher and let me know how that goes?" Think about what you'll do next. Just as when coaching teachers, coaching coaches doesn't stop. You continue until the coach's specific targets are met and you feel they're occurring at a high enough level and linked to enough natural consequences (the coach is doing it without you because it works) to move on to new targets. If you're shaping behavior, then your next steps will be finding the next level of that behavior in relation to the goal behavior.

For example, the coach is meeting with teachers more often, but the meetings are still too long and vague. The next step in this case is to focus on the quality of the coaching interactions between the coach and the teacher. Tell the coach, "If your coaching interaction for that particular instance is constructive, then why don't you follow up later that week with some positive reinforcement for the replacement behavior?" Continuity is the key to sustainable growth.

The Dichotomy of Coaching

With years of researching, leading, and supporting leadership, we've found a profound dichotomy often emerges, accentuated by the delicate balance between zooming in on specific details and zooming out to embrace a broader perspective, concepts you learned about in Chapter 10. This dichotomy lies between the coach one envisions oneself to be and the coach that the educational environment truly requires. It's a contrast that highlights the difference between personal leadership aspirations and the practical demands of effective leadership within an educational setting.

Consider the coach you might want to be, characterized by personal aspirations, goals, and preferred methods of leading. This leadership approach is deeply rooted in your perception of your ideal role and the impact you wish to have.

> For example, take Mr. Grayson, a principal who sees himself as a transformative leader, ardently focused on pioneering innovative teaching methods and integrating technology. He aspires to be seen as a progressive, tech-savvy leader steering his school toward the forefront of modern educational practices.
>
> However, the reality of Mr. Grayson's school presents a different set of challenges. The school is wrestling with fundamental issues such as improving literacy rates and supporting students from diverse socioeconomic backgrounds. The leader Mr. Grayson needs to be, in this case, is one who prioritizes these pressing needs. He must focus on inclusive education strategies and foundational skill-building, even if it means temporarily sidelining his passion for high-tech educational methods.

This shift exemplifies the transition from the leader he wants to be to the leader he needs to be, dictated not by personal preference but by the immediate and long-term needs of his students and the school.

The art of knowing when to zoom in, perhaps by concentrating on specific student groups or instructional challenges, and when to zoom out, such as considering the overall strategy for school improvement, is critical in navigating this leadership dichotomy. A skilled Deliberate Coach in this scenario must be adept at switching between the detailed focus on current, specific challenges and the broader, strategic perspective on long-term objectives and schoolwide initiatives.

This ability to fluidly transition between micro-level and macro-level focus points is what sets apart a truly effective leader in education. Such transitions require introspection, social validity measures, and responsiveness, aligning one's leadership approach with the dynamic and evolving demands of the school environment. Mastery of this balance ensures that leadership is not only aspirational but also practical and grounded in the realities of the educational landscape. It transforms aspirational leadership into a responsive and impactful force that resonates with the needs of the school community, ultimately guiding it toward sustainable success and improvement.

Complementary Perspectives: Beyond the Dichotomy of School Leadership

As you can see, being an effective leader requires a deliberate, nuanced approach, as demonstrated in the case of Mr. Grayson. However, this dynamic of adjusting focus isn't exclusive to those in formal leadership positions like principals. The same principles apply with equal importance to other key players in the educational landscape, such as consultants, instructional coaches, or professionals dedicated to supporting improved educational outcomes. These individuals, though not in traditional leadership roles, exert significant influence and are instrumental in shaping the educational experience.

> Consider an instructional coach, Ms. Khan, who works closely with teachers to enhance classroom practices. Her role requires her to zoom in on specific teaching methods, classroom interactions, and individual student needs. As a Deliberate Coach, she meticulously observes classroom sessions, provides feedback on lesson delivery, and suggests strategies for student engagement tailored to each teacher's style and classroom dynamics. In these instances, her focus is laser-sharp, attending to the minute details that can make or break the effectiveness of a lesson.

Yet Ms. Khan's responsibilities also call for her to zoom out and view these classroom dynamics in the context of broader educational goals and schoolwide initiatives. She needs to understand how each classroom's practices align with the school's curriculum standards, teaching philosophies, and overall educational objectives. For instance, while working with teachers, she considers how their approaches contribute to the school's goals for inclusive education or holistic student development. This broader perspective ensures that her coaching isn't just about improving individual classrooms in isolation but is integrated into a cohesive, schoolwide strategy.

Similarly, educational consultants, who often work with multiple schools or districts, must adeptly alternate between detailed analysis and big-picture strategizing. They dive deep into specific issues such as curriculum development, teacher training programs, or technology integration, providing tailored solutions and strategies. Yet to be effective as Deliberate Coaches, they must also step back to assess how these solutions fit into the larger educational landscape, aligning with district-wide goals, educational trends, and policy shifts.

In essence, whether one is a school leader, an instructional coach, or a consultant, the ability to seamlessly transition between a focused, detailed view and a broad, systemic perspective is essential. It's this duality of approach that enables Deliberate Coaches to effectively address immediate challenges while also contributing to the overarching mission of education. By mastering this art of zooming in and out, they ensure that their efforts are not only locally impactful but also resonate with broader educational objectives, thereby playing a pivotal role in shaping the future of education.

In summary, the journey to becoming the coach you need to be involves understanding and applying both the technical skills of zooming in and out and the philosophical shift from personal desires to serving the greater needs of your educational environment. By navigating this transition, you position yourself as a Deliberate Coach who is capable of making a significant and positive difference in the lives of your students, teachers, and the broader school community through your Deliberate Coaching toolbox.

Next Steps

Even after all these decades of research, the beauty of the science behind behavior is that we're still impacting behaviors using our understanding of basic principles, our ABCs (antecedents-behaviors-consequences). Behavior analysis can answer questions about why students are performing at their level, why teachers are teaching at their level, and even why the school district and legislators are making certain decisions.

Although not always easy, there is a way to discover why people do what they do.

The Deliberate Coaching process is one built on science and respect. Science to ensure the principles guiding teacher and coach performance are sound, and respect for those who have dedicated their lives to helping the children in our schools succeed. Training teachers is important in order to ensure they have the skills needed to effectively engage students in the learning process. Administrative support in school is critical to ensure there are processes, procedures, equipment, and standards in place to allow teachers and students to thrive. But coaching is needed to ensure long-term success (Figure 14.5).

We hope the tools and processes in this book will provide a foundation to maximize teacher performance in a way that positively impacts your school culture and student achievement. Teachers deserve recognition (positive reinforcement) for the good work they do and frequent support to help them achieve their goals as educators. Coaches deserve recognition and support for shaping teacher performance along the way. This process can't be random; it must be precise, systematic, and purposeful. It must be deliberate!

Who? Which? Why? When? Where? What? What Happened? Why Don't ... ?

Figure 14.5. Leadership is not isolated to simply selecting between administrative duties, training, or coaching (using the 8 Ws) and stopping. It is an ongoing cycle of setting people up for success and providing ongoing support.

Key Takeaways

- Accountability and encouragement start with you, the individual. Consider how you're setting *yourself* up for success, how you're holding *yourself* accountable for meeting your high standards, and how you're acknowledging the great work *you're* doing.

- The Deliberate Coaching process cascades from leader to leader, coach to coach, and peer to peer, offering levels of support and systems of positive reinforcement.

- Leadership isn't simply about coaching. Nor is it about training or setting people up for success. Leadership involves all of these facets and more.

Coaching requires training to build the skills being coached, and both leadership and coaching require systems where teachers and personnel have the tools needed to be successful and encouragement when progress is made.

- School leaders need more than individualized strategies; organizational leadership and systemic feedback are crucial for widespread student achievement.

- Instructional leadership, while valuable, may not always be a feasible first step for all leaders, especially in schools facing significant challenges.

- Individual coaching has its place but might not yield significant improvements in struggling schools without effective management and operational systems in place.

- Organizational coaching shifts the focus from individual to collective improvements, targeting systemic changes that enhance school culture and teacher performance.

- Effective organizational coaching involves simplifying expectations, standardizing procedures, and engaging staff in data-driven decision-making for schoolwide improvement.

Appendix

IMPACT Goals

INDIVIDUALIZED	
	The goal is relevant to the roles and responsibilities of the stakeholder(s).
	The goal states specifically what the desired result is.
	The goal is specific to the needs of the stakeholder(s).

MANAGEABLE	
	Staff have the knowledge and skills to reach this goal.
	Resources have been provided to support those who need to implement the change (e.g., time, tools, authority, etc.).
	Stakeholder(s) only focus on a few goals at a time.

POSITIVELY MOTIVATING	
	Stakeholder(s) understand why they should engage in the change.
	Stakeholder(s) are involved in developing the goal.
	Stakeholder(s) are motivated by the goal, or there is a plan for sustaining motivation.

ALIGNED	
	Specific behaviors and tasks required to achieve the goals have been identified.
	The behaviors and tasks identified are specific to the roles and responsibilities of the stakeholder(s).
	Achievement of this goal adds value to faculty, staff, and ultimately, the students.

CONNECTED	
	Accomplishments have been identified that can be used as salient measures of progress.
	The accomplishments are connected to the behaviors required to achieve them.
	The goal is connected to the accomplishments required to achieve it.

TRACKABLE	
	There is a process for measuring behavior required to achieve accomplishments.
	There is a process for measuring accomplishments required to achieve the goal.
	There is a process for tracking stakeholder perceptions.

Performance Diagnostic Checklist—Education

Teacher Name: _____ Interviewer: _____ Date: _____

Briefly describe performance concern:_____

_____

_____

_____

Antecedents and Information	YES	NO
1. Is there a written description stating the clear expectation of the educator regarding a particular instructional/behavioral strategy?		
2. Has the educator received adequate instruction about what to do (e.g., instructions such as "I want you to do this and this before we leave today")?		
3. Has the educator received formal training on this instructional/behavioral strategy? If yes, check all applicable training methods. ☐ Instructions ☐ Demonstration ☐ Rehearsal		
4. Is there a task aid visible *while* completing the instructional/behavioral strategy in question (e.g., reminders to prompt the strategy correctly at the right time/duration)?		
5. Can the educator state the purpose of the instructional/behavioral strategy?		
6. Is the educator verbally, textually, or electronically reminded to use the instructional/behavioral strategy? If yes, how often? ☐ Hourly ☐ Daily ☐ Weekly ☐ Monthly By whom? Check all that apply: ☐ Peer ☐ Coach ☐ Administrator ☐ Other		
7. Are there frequently updated, challenging, and attainable goals the educator is comfortable with about the instructional/behavioral strategy?		
8. Is the educator "aware" of the school's mission?		

Equipment and Processes	YES	NO
9. If equipment is required, is it available and in good working order (e.g., computer, A/V, mic, etc.)?		
10. Are the equipment and environment optimally arranged in a physical sense (e.g., the arrangement of the students' desks)?		
11. Are larger processes performing well despite incorrect instructional/behavioral strategies (e.g., routines and procedures)?		
12. Are these processes written out and arranged logically?		
13. Can the educator implement the instructional/behavioral strategy without any obstacles (e.g., interruption by the intercom)?		

Knowledge and Skills-Training	YES	NO
14. Can the educator tell you what they are supposed to be doing and how to do it?		
15. Can the educator physically/verbally precisely demonstrate the instructional/behavioral strategy?		
16. If the instructional/behavioral strategy needs to be completed quickly, can the educator perform it at the appropriate speed?		

Motivation	YES	NO
17. Are educators motivated based on the outcomes following the completion of the task?		
18. Do educators see the positive effects of implementing the instructional/behavioral strategy (e.g., increased student engagement, increased assessment data, decreased misbehavior)?		
19. Do administrators monitor the educator's behavior related to the task? If yes, how often? ☐ Hourly ☐ Daily ☐ Weekly ☐ Monthly		
20. Does the educator receive feedback about their performance? If yes, By whom? _____ and How often? ☐ Hourly ☐ Daily ☐ Weekly ☐ Monthly How long of a delay between observing the instructional/behavioral strategy and delivering feedback? _____ Check all that apply: Feedback Focus: ☐ Positive ☐ Constructive Feedback Type: ☐ Written ☐ Verbal ☐ Graphed ☐ Other		
21. Is the instructional/behavioral strategy easy to implement?		
22. Do other instructional/behavioral strategies appear to take precedence over the targeted strategy?		

Deliberate Coaching Self-Monitoring Tool (DCST)

Coachee: _____ Classroom makeup: _____ Interval: _____

Modality: _____ Coach: _____

Day/Date		Monday			Tuesday			Wednesday			Thursday			Friday		
	Scale	5	5	5	5	5	5	5	5	5	5	5	5	5	5	5
		4	4	4	4	4	4	4	4	4	4	4	4	4	4	4
		3	3	3	3	3	3	3	3	3	3	3	3	3	3	3
		2	2	2	2	2	2	2	2	2	2	2	2	2	2	2
		1	1	1	1	1	1	1	1	1	1	1	1	1	1	1
		5	5	5	5	5	5	5	5	5	5	5	5	5	5	5
		4	4	4	4	4	4	4	4	4	4	4	4	4	4	4
		3	3	3	3	3	3	3	3	3	3	3	3	3	3	3
		2	2	2	2	2	2	2	2	2	2	2	2	2	2	2
		1	1	1	1	1	1	1	1	1	1	1	1	1	1	1
		5	5	5	5	5	5	5	5	5	5	5	5	5	5	5
		4	4	4	4	4	4	4	4	4	4	4	4	4	4	4
		3	3	3	3	3	3	3	3	3	3	3	3	3	3	3
		2	2	2	2	2	2	2	2	2	2	2	2	2	2	2
		1	1	1	1	1	1	1	1	1	1	1	1	1	1	1

PINPOINT:
Defined:
Baseline:

PINPOINT:
Defined:
Baseline:

PINPOINT:
Defined:
Baseline:

Adapted from Iovannone et al. (2014).

Step-by-Step Deliberate Coaching Meeting Protocol

Step 1: INITIALIZATION (2 minutes)

- Opening Remarks: Brief overview of the meeting's importance for skill development and sharing best practices
- Agenda Overview: Outline of the meeting structure to set expectations and ensure alignment

Step 2: RANDOM COACH SELECTION (1 minute)

- Selection Method: Utilize a random selection process to identify the coach who will present

Step 3: REPORT-OUT SEQUENCE (10 minutes per coach)

- Sub-step 3.1: Establish Ground Rules: Concise, precise, specific communication to facilitate focused, constructive conversations (1 minute)
- Sub-step 3.2: Structured Sharing (leader to peers/facilitator):
  - Objective and Metrics (2 minutes): Present target results and discuss their significance for stakeholders, along with the metrics used
  - Execution Strategy (3 minutes): Discuss steps taken to meet goals
  - Results Achieved (3 minute): Describe outcomes, successes, failures, providing a complete picture of efforts
  - Future Course of Action (1 minute): Discuss next steps based on achieved outcomes

Step 4: PEER COACHING AND FEEDBACK (to the leader doing the Structured Sharing) (5 minutes)

COACH:

- C = Celebrate and Capture
- O = Observe—listen carefully and stay engaged
- A = Ask for clarity if needed
- C = Constructive—offer actionable suggestions for improvement if warranted
- H = Highlight—note down additional learnings for future application

Step 5: CLOSING REMARKS (2 minutes)

- Summary of Learnings: Recap session and emphasize actionable insights gathered
- Preview of Next Meeting: Offer a brief look ahead to foster ongoing engagement and preparation

DELIBERATE COACHING SUMMARY SHEET

Coach: _____   Month: _____

ANTECEDENTS	**Who?** Who are you coaching?				
	Why? Why is this result important?				
	Which? Pinpointed behaviors will achieve the result?				
	When? When will you coach?				
	Where? Where will you coach?				
BEHAVIOR	**What will/ did you do?** Specifically, what did you do or plan to do to coach?				
CONSEQUENCE	**What Happened?** Specifically, what impact did you have? Behavior/ outcome data				
	Why Don't … ? What will you do next?				

Coaching Self-Assessment and Checklist

Perhaps you are new to the concept of coaching, or maybe you are a seasoned pro. Either way, use this sheet to periodically check yourself against core competencies that are essential to building a culture of coaching.

Point scale: 0 = never, 1 = only when requested by employees, 2 = inconsistently, 3 = at regular intervals (once or twice a year), 4 = as often as possible.

Please rate your skills on the following questions. Then, star two areas where you feel you have strength as a coach, and check two areas where you would like to improve your coaching.

Coaching (listening and drawing out)

Talk with employees about what they like most about work (not necessarily just their current job).	
Listen to employees' concerns about the progress and direction of their careers.	
Learn what is motivating for your employees.	
Step out of the "expert" role and truly listen, just to understand your employee's perspective.	

Reality testing (providing positive and corrective feedback)

Recognize and reward employees for their contributions, in the presence of others.	
Initiate discussions with your employees about how you see their future career.	
Explain formal and informal factors that determine success in the organization.	

Advising (offering organizational insight, information, and advice)

Talk clearly about changes in the organization that will require new learning.	
Offer suggestions for opportunities for new learning.	
Work to identify on-the-job opportunities for your employees (inside and outside your immediate area) that are in line with organizational goals and employee desires.	
Provide resources for employees who want information on developmental opportunities.	

Development planning (guiding employee goal setting, following up)

Review and discuss an employee's written professional development plan and goals.	
Discuss expectations (yours and theirs) regarding professional development planning.	
Check in with employees about their progress on development goals/ initiatives.	
Provide physical support (funding, time off, personal assistance) for employees to pursue more learning.	
Provide motivational support for employees to pursue more learning.	

Rate your agreement/disagreement with each of the statements below for the following coaching pinpoints.						
COACHING PINPOINTS	**1**	**2**	**3**	**4**	**5**	**6**
This is the best coaching pinpoint that could be chosen.						
This coaching pinpoint focuses on the most important issues.						
This coaching pinpoint increases the opportunity to improve in other areas.						
This is a reasonable coaching pinpoint to accomplish.						
This coaching pinpoint will not have a negative impact.						
This coaching pinpoint will increase opportunities for reinforcement.						
This coaching pinpoint is needed more than most other pinpoints.						
1 = Strongly Disagree; 2 = Disagree; 3 = Slightly Disagree; 4 = Slightly Agree; 5 = Agree; 6 = Strongly Agree						

Rate your agreement/disagreement with each of the statements below for the following coaching goals.						
COACHING GOALS	**1**	**2**	**3**	**4**	**5**	**6**
I feel confident in my understanding of the coaching procedures.						
The coaching procedures are acceptable and ultimately linked to student outcomes.						
I am willing to follow through on the coach's recommendations.						
Given your needs and the needs of the students in my classroom, I find the coaching to be reasonable.						
The coaching I am receiving has been effective toward improving pinpointed skills.						
The coaching I am receiving is having a positive impact on pinpointed results.						
Following the coach's recommendation requires little effort from me.						
1 = Strongly Disagree; 2 = Disagree; 3 = Slightly Disagree; 4 = Slightly Agree; 5 = Agree; 6 = Strongly Agree						

References

Albergaria-Almeida, P. (2010). Classroom questioning: Teachers' perceptions and practices. *Procedia Social and Behavioral Sciences, 2*(2), 305–309. https://doi.org/10.1016/j.sbspro.2010.03.015

Albergaria-Almeida, P. (2012). Can I ask a question? The importance of classroom questioning. *Procedia Social and Behavioral Sciences, 31*, 634–638. https://doi.org/10.1016/j.sbspro.2011.12.116

Alessi, G. (1987). Generative strategies and teaching for generalization. *The Analysis of Verbal Behavior, 5*, 15–27. https://doi.org/10.1007/BF03392816

ACE Presidents' Task Force on Teacher Education. (2002). *Touching the future: Final report.* American Council on Education. https://www.acenet.edu/Documents/Touching-the-Future-Final-Report-2002.pdf

Aschner, M. J. M. (1961). Asking questions to trigger thinking. *NEA Journal, 50*(6), 44–46. https://archive.org/details/sim_todays-education_1961-09_50_6/page/44/mode/2up

Austin, J. (2000). Performance analysis and performance diagnostics. In J. Austin & J. E. Carr (Eds.), *Handbook of applied behavior analysis* (pp. 321–349). Context Press.

Austin, J., Carr, J. E., & Agnew, J. L. (1999). The need for assessment of maintaining variables in OBM. *Journal of Organizational Behavior Management, 19*(2), 59–87. https://doi.org/10.1300/J075v19n02_05

Austin, J., Weatherly, N. L., & Gravina, N. E. (2005). Using task clarification, graphic feedback, and verbal feedback to increase closing-task completion in a privately owned restaurant. *Journal of Applied Behavior Analysis, 38*(1), 117–120. https://doi.org/10.1901/jaba.2005.159-03

Badiee, F., & Kaufman, D. (2014). Effectiveness of an online simulation for teacher education. *Journal of Technology and Teacher Education, 22*(2), 167–186. https://www.learntechlib.org/primary/p/45934/

Bandura, A. (1997). *Self-efficacy: The exercise of control.* W.H. Freeman.

Banks, J., Carson, J. S., Nelson, B. L., & Nicol, D. (2010). *Discrete-event system simulation* (5th ed.). Prentice Hall.

Barnum, M. (2023, March 6). Teacher turnover hits new highs across the U.S. *Chalkbeat.* https://www.chalkbeat.org/2023/3/6/23624340/teacher-turnover-leaving-the-profession-quitting-higher-rate/

Behavior Analyst Certification Board. (2020). *Ethics code for behavior analysts.* Author. https://www.bacb.com/ethics-information/ethics-codes/

Binder, C. (1996). Behavioral fluency: Evolution of a new paradigm. *The Behavior Analyst, 19*(2), 163–197. https://doi.org/10.1007/BF03393163

Binder, C., & Watkins, C. L. (1990). Precision teaching and direct instruction: Measurably superior instructional technology in schools. *Performance Improvement Quarterly*, *3*(4), 74–96. https://doi.org/10.1111/ j.1937-8327.1990.tb00478.x

Bleiberg, J. F., & Kraft, M. A. (2023). What happened to the K–12 education labor market during COVID? The acute need for better data systems. *Education Finance and Policy*, *18*(1), 156–172. https://doi.org/10.1162/edfp_a_00391

Bosch, S., & Fuqua, R. W. (2001). Behavioral cusps: A model for selecting target behaviors. *Journal of Applied Behavior Analysis*, *34*(1), 123–125. https://doi. org/10.1901/jaba.2001.34-123

Branson, R. [@richardbranson]. (2014, March 27). *Train people well enough so they can leave, treat them well enough so they don't want to* http://virg.in/lys [Post]. X. https://twitter.com/richardbranson/ status/449220072176107520

Breaux, A. L., & Wong, H. K. (2003). *New teacher induction: How to train, support, and retain new teachers*. Harry K. Wong Publications.

Bucklin, B. R., Dickinson, A. M., & Brethower, D. M. (2000). A comparison of the effects of fluency training and accuracy training on application and retention. *Performance Improvement Quarterly*, *13*(3), 140–163. https://doi. org/10.1111/j.1937-8327.2000. tb00180.x

Cambridge University Press & Assessment. (n.d.-a). Coaching. In *Cambridge dictionary*. Retrieved April 25, 2024, from https://dictionary.cambridge.org/ us/dictionary/english/coaching

Cambridge University Press & Assessment. (n.d.-b). Training. In *Cambridge dictionary*. Retrieved April 25, 2024, from https://dictionary.cambridge.org/ us/dictionary/english/training

Carr, J. E., & Wilder, D. A. (2016). The Performance Diagnostic Checklist— Human Services: A correction. *Behavior Analysis in Practice*, *9*(1), 63. https://doi.org/10.1007/s40617-015-0099-3

Carr, J. E., Wilder, D. A., Majdalany, L., Mathisen, D., & Strain, L. A. (2013). An assessment-based solution to a human-service employee performance problem: An initial evaluation of the Performance Diagnostic Checklist— Human Services. *Behavior Analysis in Practice*, *6*(1), 16–32. https://doi. org/10.1007/BF03391789

Carter, S. L. (2010). *The social validity manual: A guide to subjective evaluation of behavior interventions* (1st ed.). Academic Press.

Cavanaugh, M. P., & Warwick, C. (2001). Questioning is an art. *Language Arts Journal of Michigan*, *17*(2), 36–38. https://doi. org/10.9707/2168-149X.1320

Chin, C. (2007). Teacher questioning in science classrooms: Approaches that stimulate productive thinking. *Journal of Research in Science Teaching*, *44*(6), 815–843. https://doi. org/10.1002/tea.20171

Chin, C., & Osborne, J. (2008). Students' questions: A potential resource for teaching and learning science. *Studies in Science Education*, *44*(1), 1–39. https://doi. org/10.1080/03057260701828101

Cooper, J. O., Heron, T. E., & Heward, W. L. (2020). *Applied behavior analysis* (3rd ed.). Pearson.

Critchfield, T., Heward, W. L., & Lerman, D. C. (2023). Fifteen years and counting: The dissemination impact of behavior analysis in practice. *Behavior Analysis in Practice, 16*(2), 399–406. https://doi.org/10.1007/s40617-022-00744-2

Cruickshank, D. R. (1988). The uses of simulations in teacher preparation: Past, present, and future. *Simulation & Games, 19*(2), 133–156. https://doi.org/10.1177/104687818801900202

Daniels, A. C., & Bailey, J. S. (2006). *Performance management: Changing behavior that drives organizational effectiveness* (4th ed.). Performance Management Publications.

Darling-Hammond, L. (2006). Assessing teacher education: The usefulness of multiple measures for assessing program outcomes. *Journal of Teacher Education, 57*(2), 120–138. https://doi.org/10.1177/0022487105283796

Davidson, Z. (2023). *An exploration of teacher attrition factors in Oregon Christian schools.* [Doctoral dissertation, George Fox University]. Digital Commons @George Fox University. https://digitalcommons.georgefox.edu/edd/212

de Bruin, A. B. H., Smits, N., Rikers, R. M. J. P., & Schmidt, H. G. (2008). Deliberate practice predicts performance over time in adolescent chess players and drop-outs: A linear mixed models analysis. *British Journal of Psychology, 99*(4), 473–497. https://doi.org/10.1348/000712608X295631

DeMatthews, D. E., Knight, D. S., & Shin, J. (2022). The principal-teacher churn: Understanding the relationship between leadership turnover and teacher attrition. *Educational Administration Quarterly, 58*(1), 76–109. https://doi.org/10.1177/0013161X211051974

Dictionary.com. (n.d.). Treatment. In *Dictionary.com.* Retrieved April 25, 2024, from https://www.dictionary.com/browse/treatment

Doran, G. T. (1981). There's a S.M.A.R.T. way to write management's goals and objectives. *Management Review, 70*(11), 35–36. https://community.mis.temple.edu/mis0855002fall2015/files/2015/10/S.M.A.R.T-Way-Management-Review.pdf

Ellis, J., & Magee, S. (2007). Lead articles: No Child Left Behind. Contingencies, macrocontingencies, and metacontingencies in current educational practices: No child left behind? *Behavior and Social Issues, 16*, 5–27. https://doi.org/10.5210/bsi.v16i1.361

Engelmann, S. (2007). *Teaching needy kids in our backward system: 42 years of trying.* ADI Press.

Erdem, F., & Aytemur, J. Ö. (2008). Mentoring—a relationship based on trust: Qualitative research. *Public Personnel Management, 37*(1), 55–66. https://doi.org/10.1177/009102600803700104

Ericsson, K. A., Krampe, R. T., & Tesch-Römer, C. (1993). The role of deliberate practice in the acquisition of expert performance. *Psychological Review, 100*(3), 363–406. https://doi.org/10.1037/0033-295X.100.3.363

Every Student Succeeds Act of 2015, 20 U.S.C. § 6301 (2015). https://congress.gov/114/plaws/publ95/PLAW-114publ95.pdf

Ferguson, R. (2014). *Finally! Performance assessment that works: Big Five Performance Management.* CreateSpace.

Fox, E. J., & Ghezzi, P. M. (2003). Effects of computer-based fluency training on concept formation. *Journal of Behavioral Education, 12*(1), 1–21. https://doi.org/10.1023/A:1022373304577

Gavoni, P. (2015, March 15). *Talent development: What leaders, managers, professional developers, and coaches can learn from boxing!* [Pulse] LinkedIn. https://www.linkedin.com/pulse/what-leaders-managers-teacher-coaches-can-learn-from-gavoni-ed-d

Gavoni, P. (2024). *Positional authority ain't leadership.* Heart & Science Consulting.

Gavoni, P., & Costa, A. (2023). *Quick wins! Using behavior science to accelerate and sustain school improvement* (2nd ed.). KeyPress Publishing.

Gawande, A. (2010). *The checklist manifesto: How to get things right.* Metropolitan Books.

Geller, E. S. (2003). Should organizational behavior management expand its content? *Journal of Organizational Behavior Management, 22*(2), 13–30. https://doi.org/10.1300/J075v22n02_03

George, A. A., Hall, G. E., & Stiegelbauer, S. M. (2006). *Measuring implementation in schools: The Stages of Concern Questionnaire.* SEDL. https://sedl.org/cbam/socq_manual_201410.pdf

Gillani, A., Dierst-Davis, R., Lee, S., Robin, L., Li, J., Glover-Kudon, R., Baker, K., & Whitton, A. (2022). Teachers' dissatisfaction during the COVID-19 pandemic: Factors contributing to a desire to leave the profession. *Frontiers in Psychology, 13.* https://doi.org/10.3389/fpsyg.2022.940718

Glenn, S. S. (1988). Contingencies and metacontingencies: Toward a synthesis of behavior analysis and cultural materialism. *The Behavior Analyst, 11*(2), 161–179. https://doi.org/10.1007/BF03392470

Glenn, S. S. (1991). Contingencies and metacontingencies: Relations among behavioral, cultural, and biological evolution. In P. A. Lamal (Ed.), *Behavioral analysis of societies and cultural practices* (pp. 39–73). Hemisphere Publishing Corporation.

Graesser, A., & Olde, B. A. (2003). How does one know whether a person understands a device? The quality of the questions the person asks when the device breaks down. *Journal of Educational Psychology, 95*(3), 524–536. https://doi.org/10.1037/0022-0663.95.3.524

Hakanen, J. J., Bakker, A. B., & Schaufeli, W. B. (2006). Burnout and work engagement among teachers. *Journal of School Psychology, 43*(6), 495–513. https://doi.org/10.1016/j.jsp.2005.11.001

Harbatkin, E., & Nguyen, T. D. (2023, October 19). *The relationship between teacher intentions, turnover behavior, and school conditions.* Brookings. https://www.brookings.edu/articles/the-relationship-between-teacher-intentions-turnover-behavior-and-school-conditions/

Harless, J. (2017). *The Eden Conspiracy: Educating for accomplished citizenship.* Cambridge Center for Behavioral Studies.

Harris, S. P., Davies, R. S., Christensen, S. S., Hanks, J., & Bowles, B. (2019). Teacher attrition: Differences in stakeholder perceptions of teacher work conditions. *Education Sciences, 9*(4), 300. https://doi.org/10.3390/educsci9040300

Hattie, J. (2009). *Visible learning: A synthesis of over 800 meta-analyses relating to achievement.* Routledge.

Heller, P. (1995). *Bad intentions: The Mike Tyson story* (Updated ed.). Da Capo Press.

Ingersoll, R., Merrill, L., & May, H. (2014, July). *What are the effects of teacher education and preparation on beginning teacher attrition?* (#RR-82). Consortium for Policy Research in Education. https://www.cpre.org/sites/default/files/researchreport/2018_prepeffects2014.pdf

Iovannone, R., Greenbaum, P. E., Wang, W., Dunlap, G., & Kincaid, D. (2014). Interrater agreement of the Individualized Behavior Rating Scale Tool. *Assessment for Effective Intervention, 39*(4), 195–207. https://doi.org/10.1177/1534508413488414

Johnson, B. M., Miltenberger, R. G., Egemo-Helm, K., Jostad, C. M., Flessner, C., & Gatheridge, B. (2005). Evaluation of behavioral skills training for teaching abduction-prevention skills to young children. *Journal of Applied Behavior Analysis, 38*(1), 67–78. https://doi.org/10.1901/jaba.2005.26-04

Johnson, D. A., & Rubin, S. (2011). Effectiveness of interactive computer-based instruction: A review of studies published between 1995 and 2007. *Journal of Organizational Behavior Management, 31*(1), 55–94. https://doi.org/10.1080/01608061.2010.541821

Johnson, D. W., & Johnson, R. T. (1998). *Learning together and alone: Cooperative, competitive, and individualistic learning* (5th ed.). Pearson.

Johnson, K. R., & Layng, T. V. (1992). Breaking the structuralist barrier: Literacy and numeracy with fluency. *American Psychologist, 47*(11), 1475–1490. https://doi.org/10.1037/0003-066X.47.11.1475

Johnson, K. R., & Layng, T. V. J. (1994). The Morningside Model of Generative Instruction. In R. Gardner, III, D. M. Sainato, J. O. Cooper, T. E. Heron, W. L. Heward, J. W. Eshleman, & T. A. Grossi (Eds.), *Behavior analysis in education: Focus on measurably superior instruction* (pp. 173–197). Brooks/Cole.

Johnson, S., Cooper, C., Cartwright, S., Donald, I., Taylor, P., & Millet, C. (2005). The experience of work-related stress across occupations. *Journal of Managerial Psychology, 20*(2), 178–187. https://doi.org/10.1108/02683940510579803

Johnston, J. M., & Pennypacker, H. S., Jr. (2009). *Strategies and tactics of behavioral research* (3rd ed.). Routledge.

Joyce, B., & Showers, B (2002). *Student achievement through staff development* (3rd ed.). ASCD.

Kagan, S., & Kagan, M. (2009). *Kagan cooperative learning*. Kagan Publishing.

Kieta, A. R. (2020, May 24). *Instruction, classroom management, precision teaching, and coaching with the Morningside Model of Generative Instruction* [Conference session]. ABAI 46th Annual Convention, Online, 2020. https://www1.abainternational.org/events/program-details/event-detail.aspx?&sid=70146&by=Domain

Klassen, R. M., & Chiu, M. M. (2010). Effects on teachers' self-efficacy and job satisfaction: Teacher gender, years of experience, and job stress. *Journal of Educational Psychology, 102*(3), 741–756. https://doi.org/10.1037/a0019237

Klem, A. M., & Connell, J. P. (2004). Relationships matter: Linking teacher support to student engagement and achievement. *Journal of School Health, 74*(7), 262–273. https://doi.org/10.1111/j.1746-1561.2004.tb08283.x

Knight, J. (Ed.). (2009). *Coaching: Approaches and perspectives.* Corwin Press.

Komaki, J., & Barnett, F. T. (1977). A behavioral approach to coaching football: Improving the play execution of the offensive backfield on a youth football team. *Journal of Applied Behavior Analysis, 10*(4), 657–664. https://doi.org/10.1901/jaba.1977.10-657

Kotter, J. P. (1996). *Leading change.* Harvard Business School Press.

Latham, G. (1997). *Behind the schoolhouse door: Eight skills every teacher should have.* Mountain Plain Regional Resource Center.

Ledoux, S. F. (2014). *Running out of time: Introducing behaviorology to help solve global problems.* Dogwise Publishing.

Levine, A. (2005). *Educating school leaders.* The Education Schools Project. http://edschools.org/reports_leaders.htm

Lindsay, J. (2014, July 10). *What the data really show: Direct instruction really works!* JeffLindsay.com. https://www.jefflindsay.com/EducData.shtml

Lippa, C. F. (2013). Loss of language skills in teachers: Is there a link to frontotemporal degeneration? *American Journal of Alzheimer's Disease & Other Dementias, 28*(6), 549–550. https://doi.org/10.1177/1533317513502251

Madsen, C. H., Jr., & Madsen, C. K. (1974). *Teaching/discipline: Behavioral principles toward a positive approach.* Allyn & Bacon.

Malott, M. E. (2003). *Paradox of organizational change: Engineering organizations with behavioral systems analysis.* Context Press.

Martin, G., & Hrycaiko, D. (1983). Effective behavioral coaching: What's it all about? *Journal of Sport & Exercise Psychology, 5*(1), 8–20. https://doi.org/10.1123/jsp.5.1.8

Martinez-Onstott, B., Wilder, D., & Sigurdsson, S. (2016). Identifying the variables contributing to at-risk performance: Initial evaluation of the Performance Diagnostic Checklist–Safety (PDC-Safety). *Journal of Organizational Behavior Management, 36*(1), 80–93. https://doi.org/10.1080/01608061.2016.1152209

Marzano, R. J., Pickering, D. J., & Pollock, J. E. (2001). *Classroom instruction that works: Research-based strategies for increasing student achievement.* ASCD.

Maxwell, J. C. (2014). *Good leaders ask great questions: Your foundation for successful leadership.* Center Street.

Megginson, D., & Clutterbuck, D. (2005). *Techniques for coaching and mentoring* (1st ed.). Elsevier Butterworth-Heinemann.

Merrett, F. E., & Musgrove, W. J. (1982). The hierarchy of reinforcement: Mainspring of the behavioural approach to teaching and learning. *Educational Psychology, 2*(3–4), 301–312. https://doi.org/10.1080/0144341820020310

Miciano, R. Z. (2004). Self-questioning and prose comprehension: A sample case of ESL reading. *Asia Pacific Education Review, 3*(2), 210–216. https://doi.org/10.1007/BF03024914

Mitgang, L. D. (2008, June). *Becoming a leader: Preparing school principals for today's schools* (Perspective Series). The Wallace Foundation. https://web.archive.org/web/20221015221221/ https://www.wallacefoundation.org/knowledge-center/Documents/Becoming-a-Leader-Preparing-Principals-for-Todays-Schools.pdf

Molenda, M., & Russell, J. D. (2006). Instruction as an intervention. In J. A. Pershing (Ed.), *Handbook of human performance technology: Principles, practices, and potential* (3rd ed., pp. 335–369). Pfeiffer.

Murphy, J., Moorman, H. N., & McCarthy, M. (2008). A framework for rebuilding initial certification and preparation programs in educational leadership: Lessons from whole-state reform initiatives. *Teachers College Record, 110*(10), 2172–2203. https://doi.org/10.1177/016146810811001002

Murphy, J., & Vriesenga, M. (2006). Research on school leadership preparation in the United States: An analysis. *School Leadership & Management, 26*(2), 183–195. https://doi.org/10.1080/13634230600589758

Naismith, J. (1996). *Basketball: Its origin and development*. University of Nebraska Press.

Neufeld, B., & Roper, D. (2003). *Coaching: A strategy for developing instructional capacity-Promises and practicalities*. The Aspen Institute Program on Education/The Annenberg Institute for School Reform. https://www.aspeninstitute.org/wp-content/uploads/files/content/docs/pubs/Coaching_NeufeldRoper.pdf

Nguyen, T. D., Lam, C. B., & Bruno, P. (2022, August). *Is there a national teacher shortage? A systematic examination of reports of teacher shortages in the United States* [EdWorkingPaper: 22-631]. Annenberg Institute at Brown University. https://doi.org/10.26300/76eq-hj32

No Child Left Behind Act of 2001, Pub. L. No. 107–110, § 115, Stat. 1425 (2002).

Ohio H. R. H. B. 512. Reg. Sess. 2017–2018 (2018). https://www.legislature.ohio.gov/legislation/132/hb512

OECD. (2020). *Education at a glance 2020: OECD indicators*. OECD Publishing. https://doi.org/10.1787/69096873-en

Patterson, B. J., Brewington, J., Krouse, A., & Hall, M. (2022). Building academic leadership capacity through coaching. *Nursing Education Perspectives, 43*(4), 222–227. https://doi.org/10.1097/01.NEP.0000000000000981

Pfitzner-Eden, F. (2016a). I feel less confident so I quit? Do true changes in teacher self-efficacy predict changes in preservice teachers' intention to quit their teaching degree? *Teaching and Teacher Education, 55*, 240–254. https://doi.org/10.1016/j.tate.2016.01.018

Pfitzner-Eden, F. (2016b). Why do I feel more confident? Bandura's sources predict preservice teachers' latent changes in teacher self-efficacy. *Frontiers in Psychology, 7*, 1486. https://doi.org/10.3389/fpsyg.2016.01486

Phillips, J. J. (2010, December 3). How executives view learning metrics. *Chief Learning Officer*. https://www.clomedia.com/2010/12/03/how-executives-view-learning-metrics/

Podolsky, A., Kini, T., Darling-Hammond, L., & Bishop, J. (2019, April 8). Strategies for attracting and retaining educators: What does the evidence say? [Special issue]. *Education Policy Analysis Archives, 27*(38), 1–43. https://doi.org/10.14507/epaa.27.3722

Reid, D. H., & Parsons, M. B. (2006). *Behavior analysis applications in developmental disabilities series: Vol. 3. Motivating human service staff: Supervisory strategies for maximizing work effort and work enjoyment* (2nd ed.). Habilitative Management Consultants.

Rodriguez, M., Wilder, D. A., Therrien, K., Wine, B., Miranti, R., Daratany, K., Salume, G., Baranovsky, G., & Rodriguez, M. (2006). Use of the Performance Diagnostic Checklist to select an intervention designed to increase the offering of promotional stamps at two sites of a restaurant franchise. *Journal of Organizational Behavior Management*, *25*(3), 17–35. https://doi.org/10.1300/J075v25n03_02

Rosales-Ruiz, J., & Baer, D. M. (1997). Behavioral cusps: A developmental and pragmatic concept for behavior analysis. *Journal of Applied Behavior Analysis*, *30*(3), 533–544. https://doi.org/10.1901/jaba.1997.30-533

Roth, W.-M. (1996). Teacher questioning in an open-inquiry learning environment: Interactions of context, content, and student responses. *Journal of Research in Science Teaching*, *33*(7), 709–736. https://doi.org/10.1002/(SICI)1098-2736(199609)33:7<709::AID-TEA2>3.0.CO;2-R

Sasson, J. R., Alvero, A. M., & Austin, J. (2006). Effects of process and human performance improvement strategies. *Journal of Organizational Behavior Management*, *26*(3), 43–78. https://doi.org/10.1300/J075v26n03_02

Schmitt, J., & deCourcy, K. (2022, December 6). *The pandemic has exacerbated a long-standing national shortage of teachers*. Economic Policy Institute. https://www.epi.org/publication/shortage-of-teachers/

Seashore Louis, K., Leithwood, K., Wahlstrom, K. L., & Anderson, S. E. (2010). *Investigating the links to improved student learning: Final report of research findings*. The Wallace Foundation. https://web.archive.org/web/20220901055645/http://www.wallacefoundation.org/knowledge-center/Documents/Investigating-the-Links-to-Improved-Student-Learning.pdf

Seniuk, H. A., Witts, B. N., Williams, W. L., & Ghezzi, P. M. (2013). On terms: Behavioral coaching. *The Behavior Analyst*, *36*(1), 167–172. https://doi.org/10.1007/BF03392301

Shapiro, E. S., & Shapiro, S. (1985). Behavioral coaching in the development of skills in track. *Behavior Modification*, *9*(2), 211–224. https://doi.org/10.1177/01454455850092005

Skinner, B. F. (1953). *Science and human behavior*. The Free Press.

Skinner, B. F. (1984). The shame of American education. *American Psychologist*, *39*(9), 947–954. https://doi.org/10.1037/0003-066X.39.9.947

Slack, M., & Oken, A. (2014, January 10). "A child's course in life should be determined not by the zip code she's born in." *Archived Obama White House Website*. https://obamawhitehouse.archives.gov/blog/2014/01/10/president-obama-child-s-course-life-should-be-determined-not-zip-code-she-s-born

Sloan, M. C. (2010). Aristotle's Nicomachean Ethics as the original locus for the septem circumstantiae. *Classical Philology*, *105*(3), 236–251. https://doi.org/10.1086/656196

Smith, T. M., & Ingersoll, R. M. (2004). What are the effects of induction and mentoring on beginning teacher turnover? *American Educational Research Journal*, *41*(3), 681–714. https://doi.org/10.3102/00028312041003681

Society for Human Resource Management. (n.d.). *Coaching in a business environment*. https://www.shrm.org/resourcesandtools/tools-and-samples/toolkits/pages/coachinginabusinessenvironment.aspx

Souza, B. (2015). *The weekly coaching conversation: A business fable about taking your team's performance—and your career—to the next level.* Productivity Drivers.

Spillane, J. P., Halverson, R., & Diamond, J. B. (2001). Investigating school leadership practice: A distributed perspective. *Educational Researcher, 30*(3), 23–28. https://doi.org/10.3102/0013189X030003023

Steiner, L., & Kowal, J. (2007, September). *Principal as instructional leader: Designing a coaching program that fits* [Issue Brief]. The Center for Comprehensive School Reform and Improvement. https://eric.ed.gov/?id=ED499255

Stürmer, K., Könings, K. D., & Seidel, T. (2013). Declarative knowledge and professional vision in teacher education: Effect of courses in teaching and learning. *British Journal of Educational Psychology, 83*(3), 467–483. https://doi.org/10.1111/j.2044-8279.2012.02075.x

Su, A. J. (2014, December 12). The questions good coaches ask. *Harvard Business Review.* https://hbr.org/2014/12/the-questions-good-coaches-ask

Taie, S., & Lewis, L. (2023). *Principal attrition and mobility. Results from the 2021–22 Principal Follow-Up Survey to the National Teacher and Principal Survey* (NCES 2023-046). U.S. Department of Education, National Center for Education Statistics. https://nces.ed.gov/pubs2023/2023046.pdf

Tompson, G. H., & Dass, P. (2000). Improving students' self-efficacy in strategic management: The relative impact of cases and simulations. *Simulation & Gaming, 31*(1), 22–41. https://doi.org/10.1177/104687810003100102

Tschannen-Moran, M., & Hoy, A. W. (2007). The differential antecedents of self-efficacy beliefs of novice and experienced teachers. *Teaching and Teacher Education, 23*(6), 944–956. https://doi.org/10.1016/j.tate.2006.05.003

U.S. Department of Education. (n.d.-a). *Every Student Succeeds Act (ESSA).* Retrieved April 26, 2024, from https://www.ed.gov/essa?src=rn

U.S. Department of Education. (n.d.-b). *Improving teacher preparation: Building on innovation.* https://www2.ed.gov/documents/teaching/teacher-prep.pdf

U.S. Department of Education. (2023a, July 27). *Education, Labor Departments announce new efforts to advance teacher preparation programs and expand Registered Apprenticeships for educators* [Press release]. https://www.ed.gov/news/press-releases/education-labor-departments-announce-new-efforts-to-advance-teacher-preparation-programs-and-expand-registered-apprenticeships-educators

U.S. Department of Education. (2023b, January 24). *Secretary Cardona announces the U.S. Department of Education's "Raise the bar: Lead the world" initiative* [Press release]. https://www.ed.gov/news/press-releases/secretary-cardona-announces-us-department-educations-"raise-bar-lead-world"-initiative

The W. Edwards Deming Institute. (n.d.). A bad system will beat a good person every time. https://deming.org/quotes/a-bad-system-will-beat-a-good-person-every-time-3/

Walker, T. (2022, February 1). Survey: Alarming number of educators may soon leave the profession. *NEA Today.* https://www.nea.org/nea-today/all-news-articles/survey-alarming-number-educators-may-soon-leave-profession

Wallis, G. (2016). "Good question": Exploring the experiences of generating questions in coaching [Special issue]. *International Journal of Evidence Based Coaching and Mentoring, S10,* 16–28. https://radar.brookes.ac.uk/radar/items/ebf84139-6c10-48aa-9476-f521158f2b7e/1/

Walsh, K. (2006, March 16). Teacher education: Coming up empty. *Fwd:, 3*(1), 1–6. https://fordhaminstitute.org/national/research/fwd-teacher-education-coming-empty

Watkins, M. D. (2013). *The first 90 days: Proven strategies for getting up to speed faster and smarter.* Harvard Business Review Press.

Weatherly, N. L. (2019). A behavioral safety model for clinical safety: Coaching for institutionalization. *Perspectives on Behavior Science, 42*(4), 973–985. https://doi.org/10.1007/s40614-019-00195-1

Wilder, D. A., Lipschultz, J. L., King, A., Driscoll, S., & Sigurdsson, S. (2018). An analysis of the commonality and type of preintervention assessment procedures in the Journal of Organizational Behavior Management (2000–2015). *Journal of Organizational Behavior Management, 38*(1), 5–17. https://doi.org/10.1080/01608061.2017.1325822

Wilen, W. W. (1991). *Questioning skills, for teachers: What research says to the teacher* (3rd ed.). National Education Association.

Wilson, C. (2007). *Best practice in performance coaching: A handbook for leaders, coaches, HR professionals, and organizations.* Kogan Page Limited.

Wolf, M. M. (1978). Social validity: The case for subjective measurement or how applied behavior analysis is finding its heart. *Journal of Applied Behavior Analysis, 11*(2), 203–214. https://doi.org/10.1901/jaba.1978.11-203

Zamarro, G., Camp, A., Fuchsman, D., & McGee, J. B. (2021, September 8). *How the pandemic has changed teachers' commitment to remaining in the classroom.* Brookings. https://www.brookings.edu/articles/how-the-pandemic-has-changed-teachers-commitment-to-remaining-in-the-classroom/

About the Authors

Paul "Paulie" Gavoni, EdD, BCBA-D

THEBEHAVIORALTOOLBOX.COM

Dr. Paul "Paulie" Gavoni, an esteemed behavior scientist, educator, and leader, has shaped human performance and organizational leadership over the span of nearly three decades. In roles such as COO, Vice President, School Administrator, and School Turnaround Manager, he's unified his innovative approach through a dedication to human behavior science, aiming to enhance performance and leadership across various sectors.

With deep knowledge of organizational behavior science, Dr. Gavoni tackles real-world problems and candidly addresses the shortcomings in performance improvement and change management. His ability to highlight practical solutions, all firmly grounded in behavioral science, emphasizes his unwavering integrity. He employs this science to foster engaging and highly productive environments, adhering to a philosophy that cultivating excellence in employees naturally leads to an environment that supports the sustainable achievement of desired results.

Dr. Gavoni's reach extends internationally across education, behavior analysis, and human service organizations. Hosting the globally acclaimed *Crisis in Education* Podcast, *Thoughts and Rants of a Behavior Scientist* Podcast, and being a *Wall Street Journal* and *USA Today* best-selling co-author, his impact resonates widely. His publications, including *The 5 Scientific Laws of Life & Leadership: Behavioral Karma*; *Quick Wins! Using Behavior Science to Accelerate and Sustain School Improvement*; and *QUICK Responses for Reducing Misbehavior and Suspensions: A Behavioral Toolbox for Classroom and School Leaders*, balance technical depth with accessibility. They provide fresh perspectives and actionable solutions to modern educational and performance challenges, all rooted in the principles of Applied Behavior Analysis (ABA).

A sought-after international speaker, Dr. Gavoni's presentation style blends professionalism with approachability, concentrating on tangible problems and realistic solutions. His authenticity and firsthand experiences facilitate a strong connection with audiences, allowing him to express concerns about educational systems and the adverse effects of certain methodologies while concurrently offering positive, science-based alternatives. Beyond his work in education and human services, Dr. Gavoni is a former Golden Gloves Champion and highly respected striking coach in combat sports. Coach "Paulie Gloves," as he is known in the Mixed Martial Arts (MMA) community, has trained world champions and UFC vets using technologies rooted in the behavioral sciences. Coach Paulie has been featured in the books *Beast: Blood, Struggle, and Dreams at the Heart of Mixed Martial Arts*; *A Fighter's Way*; the feature article "Ring to Cage: How Four Former Boxers Help Mold MMA's Finest"; FX's *The Toughman*; and the *Lifetime* reality series *Leave It to Geege*. He has also written extensively for online magazines such as *Scifighting, Last Word on Sports*, and *Bloody Elbow*, where his Fight Science series continues to bring behavioral science to MMA.

Dr. Gavoni's philosophy underscores continual advancement, positive evolution, and functional efficacy, marking him as a symbol of integrity and excellence in his field. His work not only challenges conventional thinking in education but also spans human services and combat sports, solidifying his role as a distinctive and influential figure across diverse disciplines. The multifaceted nature of his career serves as a living testament to the transformative potential of behavioral science, positioning Dr. Gavoni as a lasting voice for constructive change and a guiding force in educational renewal.

Nicholas L. Weatherly, PhD, BCBA-D

DELIBERATECOACHING.ORG

Dr. Nicholas Weatherly is a senior-level executive, leadership coach, researcher, and author with over 20 years of success leading progressive people operations. His expertise is in maximizing operating revenues and organizational and staff performance by building holistic systems and targeted training programs, linking performance-improvement initiatives to key business metrics, and strategically aligning short- and long-range goals to the organization's mission, vision, and values. Dr. Weatherly's achievements in operational excellence come through translating organizational capability and objectives into innovative organization-wide initiatives that maximize growth potential through monitoring key performance indicators, implementing science-based practices, collaborating with individuals across all levels of an organization toward a shared goal, working with interdisciplinary teams that foster diverse experiences and perspectives, and establishing results-focused objectives and timetables. He has a proven record of forecasting needs and scaling programs in fast-paced environments as market demand necessitates without sacrificing ethics or integrity.

An experienced international consultant, coach, and speaker, Dr. Weatherly maintains a line of research on leadership and coaching through Endicott College, has served as the Head of the School of Behavior Analysis at the Florida Institute of Technology, and was a management consultant with Aubrey Daniels International where he led their instructional systems projects and consulted in a number of areas of business and industry including manufacturing, energy, banking, health insurance, education, and clinical services.

Dr. Weatherly has held advisory roles and served on the board of directors for a number of professional associations, advocacy groups, and service facilities including the New York State Association for Behavior Analysis, the Minnesota Northland Association for Behavior Analysis, and the Autism Treatment Association of Minnesota. He is a Past President of the Association of Professional Behavior Analysts as well as the Georgia Association for Behavior Analysis and the Kentucky Association for Behavior Analysis. He has also worked with the Behavior Analyst Certification Board across numerous areas including serving as a member of the BACB Disciplinary Review Committee, as a Code Section Specialist for the BACB's Code Compliance Committee, as a coach trainer and mentor, and in the development of an ethics coaching system. Dr. Weatherly was the inaugural chair of the Kentucky Applied Behavior Analyst Licensing Board, one of the first stand-alone licensing boards for behavior analysis in the country, and continues to stay active in public policy efforts.

Dr. Weatherly received his PhD from Western Michigan University's Applied Behavior Analysis Program with concentrations on behavioral systems analysis, behavior-based safety, and programmed instruction. He currently serves on the Editorial Board of the *Journal of Organizational Behavior Management* and is a Board Certified Behavior Analyst-Doctoral®.

www.ingramcontent.com/pod-product-compliance
Lightning Source LLC
Chambersburg PA
CBHW071152130626
46553CB00004B/1625